WORLD'S FAIRS ON THE EVE OF WAR

WORLD'S FAIRS ON THE EVE OF WAR

SCIENCE TECHNOLOGY & MODERNITY

1937-1942

ROBERT H. KARGON
JOHNS HOPKINS UNIV.

KAREN FISS
CALIF. COLLEGE OF THE ARTS

MORRIS LOW
UNIV. OF QUEENSLAND

ARTHUR P. MOLELLA
SMITHSONIAN INSTITUTION

UNIVERSITY OF PITTSBURGH PRESS

Published by the University of Pittsburgh Press, Pittsburgh, Pa., 15260
Manufactured in the United States of America
Printed on acid-free paper
10 9 8 7 6 5 4 3 2 1

Library of Congress Cataloging-in-Publication Data

Kargon, Robert H. (Robert Hugh), author.
World's Fairs on the Eve of War: Science, Technology, and Modernity, 1937–1942 /
Robert H. Kargon (Johns Hopkins University), Karen Fiss (California College
of the Arts), Morris Low (University of Queensland), and Arthur P. Molella
(Smithsonian Institution).
    pages cm
Includes bibliographical references and index.
ISBN 978-0-8229-4444-7 (hardcover: alk. paper)
1. Exhibitions—Political aspects—History—20th century. 2. World politics—
History—20th century. 3. Science—Political aspects—History—20th century.
4. Technology—Political aspects—History—20th century. 5. World War,
1939–1945—Science. 6. World War, 1939–1945—Technology. 7. World War,
1939-1945—Propaganda. I. Fiss, Karen, author. II. Low, Morris, author. III.
Molella, Arthur P., 1944- author. IV. Title.
T395.K37 2015
607'.34—dc23   2015025619

CONTENTS

ACKNOWLEDGMENTS

THIS book began its journey with the authors' belief that the burgeoning literature on international expositions was just beginning to tap its potential as an indicator of important cultural trends in regard to science and technology. We chose to trace these trends across national boundaries, and believe that a useful, perhaps essential, way of mastering the substantial national archival and printed resources is to assemble a team of specialists who would pool their expertise, work together, and produce a coauthored book, and not a collection of separate essays. We met in person and via the Internet, produced a working session at the International Conference for the History of Technology (ICOHTEC) in Barcelona, July 2012, exchanged drafts and, with the very welcome help of Abby Collier of the University of Pittsburgh Press, delivered this book.

We owe a significant debt to many scholars, librarians, and archivists who helped us along the way. We can name only a few here, with the hope that others will accept our gratitude as well: Professor Miriam Levin, who helped launch the project; the Department of History of Science and Technology, Johns Hopkins University; librarian Chella Vaidyanathan of Johns Hopkins; and Macie Hall and the Center for Educational Resources at Johns Hopkins University. We are indebted to the Lemelson Foundation, to the Smithsonian Institution's Research Opportunity Fund for travel support in Rome, and to Dr. Carlotta Darò and Dr. Mariapina di Simone for their assistance in navigating the Archivio centrale dello stato in EUR. In Australia, we thank the School of Historical and Philosophical Inquiry at the University of Queensland. We would like to thank the DAAD for supporting research in Germany, as well as Iris Lauterbach and Christian Führmeister at the Zentralinstitut für Kunstgeschichte, Munich, for sharing their scholarly expertise. Gratitude is extended as well to Dr. Benedikt Mauer for his generous assistance at the Stadtarchiv Düsseldorf, and to

James van Dyke for sharing his knowledge of the *Schaffendes Volk* archives. Travel assistance from the California College of the Arts, the Smithsonian Institution, Johns Hopkins University, and the University of Queensland made our Barcelona working session at ICOHTEC possible.

WORLD'S FAIRS ON THE EVE OF WAR

ONE

INTRODUCTION

World's Fairs, Modernity, and the Demand for Authenticity

THIS book is a comparative study of how five nations during the tumultuous 1930s engaged in a fierce ideological struggle to define the future. The venues were world's fairs; the means were dramatic displays of their own national versions of modernity. The international expositions planned or mounted just before the outbreak of the Second World War are especially revealing. These expositions reflected the political regimes of the host countries, and in some cases serious divisions within them. They also highlight increasingly tense ideological divisions among nations representing liberal or social democratic republics (France and the United States), communist government (the Soviet Union), and reactionary modernist or fascist regimes (Germany, Italy, and Japan).

This book will examine world's fairs and expositions that were extensively planned just before the outbreak of hostilities in 1939, drawing upon three actually built—Paris 1937, Düsseldorf 1937, and New York

1939—and two planned in detail but never executed, Tokyo 1940 and Rome 1942. The chapters will illuminate the representation of science and technology at these fairs as indicators of modernity as part of the ongoing culture and propaganda wars preceding the outbreak of one of the most horrific conflicts of modern times.

These expositions and fairs differ from their predecessors in one fundamental way: they focused their spotlights on ideological struggle. The first fair to do so was Paris 1937. As the *New York Times*'s foreign correspondent Anne O'Hare McCormick noted at the time, the traditional world's fair celebrations of patriotism were being replaced by creeds such as fascism, Nazism, and communism, and their claims to the future.[1] A fuller understanding of what McCormick termed "national projections" at these major events requires three essentials: an inspection of the evolving role of world's fairs and expositions, an examination of international relations during the run-up to the expositions, and an appreciation of the internal ideological situation in each nation.

The First World War challenged many common assumptions inherited from the nineteenth century, especially about unlimited progress, the role of technology, growth, and not least, the naturalness of the political order. The dynastic empires—Russia, Germany, Austria-Hungary, the Ottoman Empire—which had seemed so solid, just melted away. The thrones that had held together disparate peoples quickly disappeared. British rule in Ireland and the monarchy in Italy teetered. Class warfare became more intense, even in the relatively liberal states of the West. In a world flying apart, new leaders intensified their search for unifying, centripetal forces. For many, the bright, shiny promises of science and technology required reevaluation. For some, the loss of recognizable common values and common goals was at the core of the crisis. For them, modernity was conceived as a machine without a soul. In Western Europe in the 1920s and 1930s, America—especially the skyscrapers of New York and Fordist mass production—became emblematic of modernization without purpose, without heart or soul.[2] What was required was to infuse spirit into the machine. Authenticity—what makes us who we are—and how to vitalize modernity became searing questions.

Such issues were sorted out in a number of different ways, and led to significant national varieties of modernity. Several authors have examined how public cultures were constructed in the late nineteenth and twentieth centuries. Eric Hobsbawm, for example, notes how schools, public ceremonies, expositions, and public monuments were enlisted in molding public sentiment by what he terms the "invention of tradition."[3] Maurice Roche has discussed the role of "mega-events" such as world's fairs, expositions, and Olympic Games in the formation of public cultures.[4] Paul Greenhalgh points out that "throughout the nineteenth and twentieth centuries, [world's fairs] were the only events capable of bringing such a wide selection of people to the same place for the purpose of edification and entertainment. They were intended to distract, indoctrinate and unify a population."[5] Thus, along with radio, cinema, print media, monumental architecture, staged public events, and elaborate funerals of public figures, world's fairs and international expositions were to play a major role in defining and displaying the various national versions of modernity.

World's fairs in their now recognizable form began with London's Crystal Palace Exposition of 1851. These fairs were products of the mature Industrial Revolution, an upheaval that saw the birth of new social classes, the creation of massive cities out of old market towns, and political upheavals in Europe and North America. In short, the Industrial Revolution marked a new world being born. The world's fairs, or international expositions, were part of an attempt to bring a semblance of order into this world that often seemed about to lurch out of control.

As early as 1848 Marx and Engels wrote in *The Communist Manifesto* of modernity in the form of industrial capitalism as a centrifugal force. They marveled at its "constantly revolutionizing the relations of production and with them the whole relations of society." They described "all fixed, fast frozen relations" being swept away. "All new formed ones [become] antiquated before they ossify. All that is solid melts into air. All that is holy is profaned."[6] Examining the other side of the coin, Émile Durkheim, the French sociologist, concerned himself with the need for social solidarity in a complex industrial society. For Durkheim, organic solidarity (as he termed it) was fortified by the mutual interdependence demanded by specialization and the division of labor. But building what Durkheim termed "collective consciousness" (*conscience collective*), the totality of which could serve as a solidifying force, remained a problem for modernizing societies.

Expressing and to a large extent shaping collective consciousness are society's institutions of communication and education such as schools, museums, and mass media. In this light, world's fairs or international expositions played a new and important role beginning in the last half of the nineteenth century. The world's fairs became a way for national and local economic and political elites to educate the populace by encouraging a collective consciousness more welcoming to the vast changes then being experienced.[7] They made these novelties seem less threatening by relabeling rapid change as "progress." They portrayed these transformations as natural and indeed inevitable. If modernity had its costs, these costs, if mentioned at all, were displayed as minor.

Beginning with London's Great Exhibition of 1851, world's fairs were exercises in mastering the Industrial Revolution. They celebrated national skills by displaying inventiveness, the production and distribution of goods, and advances in communications and transport, and at the same time placing all these innovations within a comprehensible and comforting context of history, tradition, and high art. The depiction of past and future at these popular events enabled a benign view of what must have seemed at the time incomprehensibly rapid alterations. In short, world's fairs encouraged a culture disposed to accept change itself as a positive good.

By the end of the century and the uniting of laboratory science and technology, leaders in science-related industries such as the electrical and chemical industries used world's fairs to come together to set standards to nurture the growth of the industries and benefit the wealth of nations. Also by the end of the nineteenth century, land grabs in Asia and Africa allowed nations participating in world's fairs to display their command of modernity (increasingly defined in industrial, military, and bureaucratic terms) by showcasing conquered "primitive" peoples and their folk arts and crafts. These "colonial" displays served to underline the necessity of modernization and the costs of nonparticipation. They served one other important purpose: to justify the colonial mission as one of bringing primitive peoples within the compass of modern civilization.

Accordingly, by the beginning of the twentieth century, world's fairs began to assume a somewhat different coloration. Instead of concentrating mainly on displays of industrial prowess for purposes of trade, they began to emphasize national economic, military, and scientific might. The major powers began to stress their claims to a dominant role in shaping the future, and the minor powers their claims to a seat at the table.

By the 1930s and the advent of the worldwide Great Depression, there existed a basic consensus regarding the foundation of what was termed "the modern. " All nations participating in the international expositions represented in this volume accepted the necessity of industrialism and the importance of science-based or "high" technology. All understood the inevitability of the rationalization of production despite differing interpretations regarding its implementation. All grew their own bureaucracies and most recognized the importance of rational, strategic economic and social planning. The critical differences among them centered upon how each nation was to come to terms with modernity with regard to what it deemed its own national character. Each in its own way chose to present a unifying and inspiring message to its people and to the world in order to display a posture that demanded respect and, in some cases, fear.

Fear has consequences. In 1931, Japan invaded China and subsequently assumed control over Manchuria. The League of Nations named Japan the aggressor; Japan withdrew from the League and distanced itself from Western allies. The United States opposed Japanese conquests in China and by the end of the decade imposed export restrictions. In 1935–1936 Italy invaded Ethiopia in order to subjugate it and reduce it to colonial status. The League of Nations imposed relatively weak sanctions on Italy, and failed to impose, after much discussion, oil and gas sanctions.[8] In 1936 Germany remilitarized the Rhineland in contravention of the Treaty of Versailles. Only the Soviet Union urged sanctions at the League of Nations; Great Britain refused to consider them. Demanding appeasement, "the dynamisms" (as McCormick phrased it) of the authoritarian nations at world's fairs undeniably had important international consequences.

For Italy, its depiction of the ancient Roman Empire served as both justification and prophecy for Mussolini's new imperium. Just as the old empire provided important roots of European civilization, the new Italian empire would lead through science, technology, art, and culture, and once again dominate the Mediterranean world. For a racially and culturally united Germany, the Aryan nation would satisfy its destiny through scientific and technical modernity while demonstrating its unity and power through its timeless roots in blood and soil. The Japanese empire, united by kinship and history through their divine emperor, aimed at exerting pan-Asian leadership through its mastery of Western technology infused with its Asian spirit and values.

The two democracies represented in this volume, France and the United States of America, mounted expositions whose multiple messages reflected the political processes that produced them. France acknowledged and celebrated its diversity—of geography, of industries and agriculture, of types of workers—as well as its commitment to democracy. But in a world of mass production, France's uniqueness lay in the application of its intelligence and skills in design. Blending art and technology, France could make sense of and add value to modernity. America's fair was an open, though muted, contest between two factions: those who saw the fair as an opportunity to educate the public in its power and responsibility to plan and create a new and better world, and those who wished to kick-start the Depression-era economy by portraying a future created by corporate research and American individualism.

The following chapters examine these pathbreaking international expositions mounted or planned by nations that would soon be at war. Two of the fairs discussed, though extensively planned, were never opened —Tokyo 1940 and Rome 1942. The time frame of 1937 through 1942 marked supremely intense ideological rivalry, for these five nations were competing not merely for industrial leadership but for command of the world of the future. The struggle that would, in a few short years and months, be waged by armies and navies was then (when the fairs were in their planning stages) still, for the most part, propaganda warfare. As it turns out, the war of ideas was not only remarkably revealing but significant for the outcome of the fighting as well.

TWO

MODERNITY À LA FRANÇAISE

The 1937 Paris Exposition

AS disorderly as the European politics in which it was embedded, the 1937 Exposition internationale des arts et techniques dans la vie moderne opened officially in Paris toward the end of May. It was only partially completed and no one was certain about its final readiness. About one theme there was a broad consensus: France had to find its niche, its own version of France's special role in a rapidly changing world. But precisely what that niche was to be produced a variety of responses. This chapter focuses upon *technique*, its relations to arts, and its role in defining French modernity at the 1937 Exposition.

The fairgrounds extended from the Champs de Mars to the new Palais de Chaillot across the Seine. When completed, the exposition boasted about three hundred pavilions, the most imposing of which were the enormous German and Soviet edifices.[1] It was attended by over twenty-four million visitors and despite strikes, delays, high costs, carping criticism from the political left, right, and center, and complaints from aesthetic critics of all stripes, the exposition was generally viewed as a success in the popular press.[2]

The Roots of the Exposition

The exposition of *arts et techniques dans la vie moderne* had grown out of a series of proposals dating back to 1928. The initial government proposal of 1928 set out to replicate the critical and popular success of the Exposition des arts décoratifs et industriels modernes of 1925, the theme of which had been the superiority of French-designed luxury manufactured goods ranging from textiles to automobiles. Its goal was to confirm France's role as the global leader in good taste and fashion. By this time, world's fairs were increasingly disengaging from their former charge to disseminate highly technical information; the 1928 proposal sought to make an *exposition universelle* by including other nations and focusing on consumers rather than manufacturers.

A major alteration in the planning for the 1937 Exposition occurred, however, owing to the riots over accusations of government corruption (the Stavisky affair) in February 1934. In a matter of days the centrist cabinets of the premiers Chautemps and Daladier fell, and the existing plans for the 1937 world's fair at Paris were canceled. As soon as former commerce minister Gaston Doumergue was named premier and formed a center-right government, he was besieged from all sides with calls for reinstatement of the 1937 Exposition effort. Doumergue responded by dropping virtually all of the previous plans and the prior leadership. He replaced them with nonpoliticians, chief among them his friend the long-time civil servant Edmond Labbé, whom he appointed to the post of commissioner general. As Labbé's first assistant he named Paul Léon, a professor and architectural historian at the Collège de France. The plans for the exposition were reframed and reformulated. Under the center-right government and the pressure of economic crises, the exposition committee reworked the proposal to add a section on rural life and agriculture—still large sectors of the French population and economy—that were funded and approved later under the new Popular Front government.[3]

Labbé was the son of artisans who began his career as a teacher and rose in the Ministry of Education to become in 1920 the director general of technical education of France, a post he held until his mandatory retirement in 1933. As director general, Labbé rejected the sharp division between the classical curriculum, stressing ancient languages and literature, and technical education. He encouraged and tried to advance what he termed *les humanités techniques*, an approach he envisioned as bridging the cultural gap between artisan and French high culture, as well as healing social rifts between classes.[4] In the late 1920s and early 1930s, a significant number of writers evinced a broader philosophical concern with the establishment of a *culture technique* that would meet the needs of an industrial society and link to a broader general culture. These writers included the philosopher-journalist Alain [Emile Chartier], Labbé's friend the poet Paul Valéry, and the political writer-geographer André Siegfried.[5] In his article "Culture and Technique," translated for the *Harvard Business Review* in 1934, Roger Dautry saw *culture technique* as a middle way between the extremes of one-sided technical education ("technique before all") and an exclusive "culture above all" approach. He lauded Labbé:

> The admirable Directeur Général de l'Enseignement Technique, M. Labbé, understands this perfectly. He has not confined technical education within the limits of the mere pursuit of the calling, nor has he reduced it to a mere apprenticeship for a single trade nor isolated it from general education. Side by side with each phase of the latter he has placed a phase of professional education. The combination of the two systems offers to all men . . . that solid foundation of *humanités primaires* on which the future edifice of their lives may rise to greater heights.[6]

Labbé's *humanités techniques* promised a powerful response to the challenge of what was perceived to be the American version of modernity: the implications of Fordist mass production and consumption, that is, the rationalization of entire production-consumption chains. As elsewhere in Europe, in France the 1920s and 1930s produced a spate of authors providing analyses of American modernity and comparisons with their own country. Roger Gagnon's still valuable analysis and review of the French discussion has been augmented in recent years by considerable scholarly attention to the issue.[7] No discussion of the France-America debate in the 1930s, however, can avoid the wild popularity both in France and in the United States of Georges Duhamel's vitriolic *Scènes de la vie future* (1930) and its more appropriately titled, because of its condemnatory tone, English version, *America the Menace* (1931). Though trained as a surgeon, Duhamel was the winner of the Prix Goncourt (a prize for French literature) for 1918 and was elected to the Académie française (the official authority on the French language) in 1935. Duhamel was an avowed antimodernist, rejecting what he saw as the mechanization of the spirit. For him, America was the chief culprit in its addiction to the "excesses of industrial civilization." Its people were "caught in the meshes of a machine of which soon no one will know the secrets."[8]

With a more moderate tone and a veneer of constructive criticism, André Siegfried, in his *America Comes of Age: A French Analysis* (1927), was in most readings openly contemptuous of American civilization and its new constituents, especially the Jews, who included "verminous refugees from the ghettos" and who were "pseudo-American[s]" who "have invaded their hotels until they have crowded the Gentiles out."[9] Like other commentators, Siegfried decried the American commitment to mass production that in his eyes was "quite unsuited to a whole group of industries in which the extreme development of machinery is . . . detrimental to the creative genius on which they depend. We now enter an entirely different field where French workmanship is the criterion." Siegfried continued:

> Here the value is in the originality and finish of the article and competition requires not the standardization of a few types, but a higher level of perfection and a greater variety of models. The spirit of this kind of industry is in direct opposition to the American system, for mass production destroys the value of an article where distinction is the main attraction. . . . French strength and invulnerability lie in creative instinct and the quality and individuality of the product. The French are in fact stronger when labour is less mechanical and more personal and artistic.[10]

These views on mass production resonated with Labbé's early ideas, now reinforced by the volume and tone of the contemporary modernization discourse. Labbé's commitment to a French answer to the standardization of mass production resurfaced during his tenure as commissioner general (general manager) of the 1937 Paris Fair. As Shanny Peer has pointed out, Labbé favored two themes that Peer explores in depth in her book, *France on Display*: regionalism and the role of artisanship. The latter was a position bolstered politically by the participation of "quasi-corporatist organizations defending French artists and artisans in the 1930s, which were influential on the fair's planning commission."[11] Labbé himself underscored his differences from America when, with his tongue firmly planted in his cheek, he extended an official invitation to the exposition to Sinclair Lewis's fictional George Babbitt.[12]

The official monthly magazine published by the General Committee, the English-language version titled *Exposition Paris 1937: Arts, Crafts, Sciences in Modern Life*, appeared in May 1936 and was printed through August 1937. The first issue clearly lays out the commissioners' goals and reflects Labbé's viewpoint: "Great men have given their lives to Thought. Lesser men have dedicated their all to work of the hands. Is it not possible to make an amalgamation? We hope for a new aristocracy—an aristocracy that will include Thinker, Scientist, Craftsman, Artist, Worker. This is the true aristocracy of twentieth-century progress."[13]

France would emerge as the aristocrat of modernity. As Paul Léon, assistant general commissioner of the exposition, explained, the exposition was designed to position France as offering a third way between two socioeconomic extremes, "Asiatic" subsistence production and "American production which is equipped and standardized to an extent we cannot even imagine." The solution was an exposition that showcased a way of life that was not highly mechanized, but dignified the role of human creativity in bringing science and technology to heel. "Whatever the problems under consideration," he urged, "they are always presented under the sign and in the setting of artistic invention." Quoting the eighteenth-century French statesman Jacques Necker, he concluded, "Good taste is the most profitable French industry."[14]

In his final report Commissioner Labbé echoed the assistant commissioner's words, while emphasizing that creative people are simply special kinds of workers. Declaring the solution was an exposition that "has emphasized the work of man and not that of the machine,"[15] Labbé illuminated his view of progress by linking it to innovative societies that the French identified with their own history and the history of Western civilization. The wedding of art and technology in social practices of all kinds, he assured readers, had venerable roots in ancient Greece, which had coined the term *techne*. As for "métiers," alluding to the great innovations of French gothic culture, Labbé described these contemporary specialties as adaptations evolved from medieval ways of organizing labor. The present does not strive to revive the past, but retains some essential characteristics useful for advancement, especially for building unity and economic growth at home. Education is fundamental; it always has been an integral part of republicanism. The exposition intended not only to celebrate and highlight these values but also to educate the public to appreciate and respond to them. Its exhibits displayed republican faith in human rationality and the utility of science and technology. The end result, Labbé promised, was "to stimulate French production . . . pull inactive Capital out of its torpor . . . and contribute to reviving the economic vitality of France."[16]

Exposition Plan

The exposition planners organized it into an elaborate taxonomy with fourteen groups and seventy-five classes. Group I (Thought Conveying), for example, had as its president Paul Valéry of the Académie française, and was divided into seven classes. Class 1, Applied Scientific Discovery, was headed by Jean Perrin, Nobelist and member of the Institut de France. Class 2, Libraries and Literature, was headed by Georges Duhamel, the aforementioned author of *America the Menace* and member of the Académie française. Class 5, Ballets and Music, had as its president the well-known composer Albert Roussel. Class 6, Cinema, had as its president the film pioneer Louis Lumière.[17] This complex organization was structured in order to evince an inner unity: "It is hoped that this Exposition will form a general synthesis of the highest form of art and science." The Craft Center would enable the visitor "to see how things are made, from the moment the artist has his inspiration, until the scientific craftsman interprets the dream." The Regional Center would demonstrate that "though widely separated by localities, manners and language, the people of France are One when it is a question of progress." The Festival of Light to be held on the River Seine was meant to symbolize the entire exposition: "The symbolism behind it is of the utmost importance: the uniting of art, craft and science."[18]

In sum, the plan for the exposition rested on a larger idea about the true basis for the organization, character, and calling of French society: the ideals on which the Third Republic was founded. The exposition was the material embodiment of a particularly French republican cultural system where an environment of open exchange (politically, socially, and economically) encouraged wedding human creativity with science and technology among all French people uniting them for their own and others' benefit. They attributed the values of liberty, equality, and fraternity to all French people and found in science a common rational foundation for their development. The emphasis on *les arts et techniques dans la vie moderne* was an opportunity to make these abstractions manifest in the context of contemporary life. Tapping into these inherent national sentiments seemed to offer a cure for France's political and economic problems and place her once again at the head of world progress.

In June 1936 Léon Blum and his democratic socialist Popular Front government assumed power. Planning for the exposition was already well underway, although actual construction was not. Blum actively engaged with plans for the exposition, giving them what is generally thought to be a modernist twist. Though he kept Labbé as commissioner general, he added his own liaisons, especially Max Hymans, the undersecretary of state for Commerce, who oversaw fair operations, greatly expanded the budget, and added new pavilions under the direct supervision of government ministries. The new pavilions included those concerning aeronautics, public works, a rural center by the Ministry of Agriculture, hygiene, and "Solidarity." He approved a Work Pavilion mounted by the General Confederation of Labor (CGT), and Le Corbusier's controversial Pavillon des Temps Nouveaux.[19]

Labbé's emphasis on regionalism and artisanship was not, to be sure, antithetical to the Popular Front's program. Both could fit well within the government's desire to open up the heights of French culture to all social classes. Jean Perrin's Palais de la Découverte is a good example. Perrin was a Nobel laureate and the undersecretary of state for scientific research in the Blum government. Along with physicist Paul Langevin, he conceived and planned a massive science center, ultimately housed in the Grand Palais.[20] In the June 1936 issue of *Exposition Paris 1937* Perrin promised that "by means of moving pictures, sound films, [and] actual experiments performed in public before the eyes of the world, all the richness of scientific progress will be explained to those who come to the Palace of Discovery . . . Those who have the urge for research should be encouraged whether they be rich or poor, aristocrat or proletariat."[21] A year later, in the May

Modernity *à la française* 13

1937 issue, Perrin lauded Commissioner Labbé and Assistant Commissioner Léon for their support: "They energetically took the rather daring responsibility of obtaining the funds and using them to this end. . . . In an Exposition devoted to Arts and Crafts in Modern Life they determined to give their rightful place to those activities which are the fountainhead of all those crafts and methods, and hence of all our power."[22] The Palais de la Découverte was one of the most popular exhibitions at the fair.[23] In the mind of Jean Perrin, its organizer, the Palais de la Découverte was a testimony to the power of the premier group of *gens de métiers*: the scientists. An ardent republican socialist, Perrin considered scientists (himself among them) to be research workers. These were men whose disinterested work made them the fundamental social benefactors. Their lack of concern for politics, economics, or utility as they worked had opened a vast new world made manifest in the exposition itself: "Without them no such exposition [on modern life] could take place." In the English version of the official magazine in June 1936, he wrote, "We who live in every-day life profit more than we realize by the scientific obsession of research workers."[24]

To make the point that there was a "harmony between discovery and application" Perrin had divided the space into a large central area devoted to explaining a series of great scientific discoveries as creative thought processes, while a wing of the Palais housed the history of important applications of these discoveries to health, industry, and commerce. Since the foundational center of the exposition was science and arts, the Palais de la Découverte aspired to be its crown. The center stretched from a main entry at the Grand Palais along the Seine to the Trocadero. As undersecretary for scientific research in the Blum government and a member of the international Confederation of Intellectual Workers (CTI), Perrin described the agenda, which focused displays on scientists as creative workers, their discoveries, their process of discovery, and—in a side wing—some of the important uses to which scientists' basic research had been put.

The Palais was a logical foundation on which the rest of the exposition rested, for it showed scientists as single-minded, working free of political economic and utilitarian pressures. They opened a vast new world that was on view. While there was certainly a celebratory side to the Palais's exhibits, Perrin had worked to marshal a variety of display techniques, in-

Fig. 2.1 Viewing Léger's *Power Transmission* mural at the Palace of Discovery. Album officiel: Exposition internationale des Arts et des techniques appliqués à la vie moderne—Paris 1937

cluding photographs, instruments, demonstrations, and interactive electric and mechanical displays, to enable all visitors to see progress. With "pure" or basic scientific discoveries in the main space of the Grand Palais and the "applied" discoveries in a wing to the right, Perrin felt that the public would see the "harmony between discovery and application: art and thought." In May 1937 Perrin described his soon-to-be-opened Palace of Discovery in this fashion: "For the first time the general public will have an opportunity to realize how predominant a part in the development of Civilization has been played by the discovery of the Unknown. Everyone will understand that discoveries must continue unhampered by material ambition and practical considerations if we would achieve in the future results comparable to the past." Perrin emphasized the democratic character of the Palais at the disposal of and for all. First, his aim was to educate, to produce in many a general understanding and in all an appreciation; the second aim was to encourage young people from all sectors of society to become scientists; the third aim was political. He wished to further his goal of establishing a French foundation for funding research laboratories and scientific schools.[25]

From the start the organizing committee of the fair, the Superior Council, had a substantial representation of artists, especially the decorative arts and including representatives of the Union corporative de l'art français.[26] Moreover, whether as an act of repaying a political debt or of carrying out its core beliefs, the Blum government incorporated modern art into the scientific and technical pavilions and celebrated modern modes of communication, transportation, and production. The Palais de la Découverte prominently displayed modernist art, including *La houille et ses dérivés* by André Lhote, and the mural *Transport des Forces*, or *Power Transmission*, by Fernand Léger and his students. The latter piece exalts the powers of nature as represented by a luxuriant waterfall and rainbow, along with the strength of technology represented by elements of an electric power station.[27] The other science and technology pavilions were also well attended and most of them also reflected a modernist flavor. One of the most striking was the Palais de l'Air, with its huge glass canopy suggesting the nose of a giant airplane. Inside was suspended a large airplane surrounded by orbs designed by Robert Delaunay and large panels in the great hall by both Robert and Sonia Delaunay.

The radio pavilion boasted demonstrations of television and during the Festival of Light broadcast the music of Darius Milhaud and Jacques Ibert.[28] The Palais de la Lumière et de l'Electricité, built by Robert Mallet-Stevens, on the Champs de Mars sported an enormous painting by Raoul Dufy, ten meters high by sixty meters long and titled "The Electric Fairy."[29] The painting projects two themes, first the history and second the applications of electricity. The lower part portrays over a hundred inventors and scientists who contributed to the development of electrical technology.[30]

The Palais de Chemins de Fer was notable for its façade marked by five translucent columns before four huge decorative panels by Félix Aublet and Robert Delaunay. Both Robert Delaunay (*Symphonie ferroviaire*) and Sonia Delaunay (*Eau*) contributed paintings.[31]

Fig. 2.2 The Palais de l'Air. Album officiel: Exposition internationale des Arts et des techniques appliqués à la vie moderne—Paris 1937

Besides illustrating the march of science and technology with fine art, the exposition used new high-technology communication as an exemplar of the union of science and culture and of the democratization of high culture. In July, some months after the opening of the exposition, the French office of Post, Telephone and Telecommunications (PTT), began broadcasting via television dance and music performances, theatrical performances, and interviews. At the fair itself, in September, the PTT broadcast a one-act play, "Evocations. Mozart enfant."[32] Cinema, long presented as a French art/technology, was heavily featured at the exposition. First displayed at the 1900 Paris World's Fair, films played a significant educational role at the 1937 Exposition. The Photo-Ciné-Phono Pavilion near the Eiffel Tower paid homage to the pioneers of French cinema, showed how films were made, and boasted a thousand-seat theater. Hundreds of documentary films were shown at over forty theaters throughout the fair including the pavilions for education, hygiene, aluminum, tourism, and railroads, the Rural Center, the Regional Center, and many international pavilions. For Labbé, film provided a powerful means to advance his themes: "[Cinema] will bring us visions of the world outside, of pro-

Modernity à la française 17

vincial life, of regional professions. It will contribute to the knowledge of craftsmanship the world over. It will present the faces of our country to our visitors."[33]

The Foreign Pavilions: War by Other Means

The *New York Times*'s foreign correspondent, Anne O'Hare McCormick, perceptively saw the 1937 Paris Exposition as marking a sea change in the character of world's fairs. She told her readers that, "For the first time so blatantly, the national pavilions are conceived and executed as 'national projections.' Today a world's fair is not an exposition of national arts and industries. It is not even a commercial show to encourage trade. . . . [I]t is a display of one thing to prove another." What the nations were trying to prove were "efficiency, ingenuity and power," and consequently the exhibits became "almost war-like." She goes on to conclude, "If for no other reason, the Paris Fair is interesting as the first exposition of this interesting new flaunting of political parties and symbols as distinct from nations in the old sense. Itself an enterprise of the French Popular Front, its most conspicuous foreign exhibitors are Sovietism, National Socialism and Fascism decidedly as such. Apparently, the democracies are overshadowed by these dynamisms."[34]

The French way of the modern—a path illuminated, in the view of the exposition planners, by intellect, taste, and skill—seemed pallid before the muscular and full-bodied versions of the ideologically driven. The Soviet Union and Nazi Germany, ideological adversaries, dramatically illustrated their opposition through their huge, imposing pavilions. Situated opposite one another across the central axis of the exposition, and developed in secret competition with one another, the two attracted worldwide attention then and have been the subject of considerable analysis since.[35] These two gold-medal-winning pavilions were by far the most popular foreign exhibits at the exposition, with Italy a distant third.[36] The Soviet pavilion, designed by Boris Iofan, was essentially a thrusting, skyscraper-like platform for a six-story statue by Vera Mukhina of a male worker and a female worker holding aloft a hammer and sickle. Inside, using electrified maps (inlaid with jewels) and diagrams as well as murals, the exhibition aimed at presenting a portrait of the Soviet Union as a territory of vast natural and human resources, rapidly industrializing and modernizing. New applica-

Fig. 2.3 The Soviet Pavilion 1937. Album officiel: Exposition internationale des Arts et des techniques appliqués à la vie moderne—Paris 1937

tions of technology received considerable attention: air travel, a Moscow metro car, an iron and steel works. Prominently displayed as an avatar of modernity was a huge black Soviet Ford limousine, branded as the GAZ, halfway up the central stairway.[37]

Just as the Soviet pavilion aimed at projecting size and material and social progress, the German pavilion aimed at projecting Germany's strength—economic, technological, and spiritual.[38] Albert Speer's updated classicism of the exterior envelope masked the innovative steel structure.[39] Inside, in an ambience of bourgeois comfort and grand chandeliers, the Germans displayed their most modern technologies: industrial machines, a swastika-branded Mercedes-Benz automobile, video-telephone, television, and a red synthetic rubber (Buna) floor, perhaps echoing the field of the Nazi flag. As Karen Fiss has written, the exhibition "tried to market Nazi Germany as both tied to its timeless roots and technologically superior and future-looking."[40]

Italy had at first declined to participate in the fair because of sanctions imposed by the League of Nations after its Ethiopian invasion. But the Fascist regime soon shifted course, deciding to use the fair as a propaganda opportunity, a triumphant announcement of its recent conquest of Addis Ababa and Mussolini's ensuing Declaration of Empire.[41] In fact the Paris exposition became part of Italy's propaganda strategy built around international exhibitions, including the Italian pavilion at the 1939 New York World's Fair and, more importantly, the grand but aborted Esposizione universale di Roma (E42) in 1942, the subject of a subsequent chapter.

While Mussolini was cementing his ties with Hitler in his bid for world-power status and aimed to make E42 the greatest fair of all time, his Italian pavilion in Paris, inaugurated in May 1937, was a relatively modest affair, staying on the sidelines in the infamous faceoff between the bombastic Soviet and Nazi buildings. Nevertheless, it revealed the high stakes and complexities of Fascist cultural politics in the prewar years. Central to this was the Fascist attempt to redefine Italian identity in a way that blended modernism with tradition. As floridly phrased in Edmond Labbé's official report on the Italian pavilion: "The goal of Italian participation was to put on display . . . the harmonization between the most audacious modernism with the most profound respect for the formidable patrimony of its traditions."[42] In so doing, Labbé pointed out, Italy epitomized the "gift" to the world of the "races méditerranéennes." He clearly embraced Italy's contribution, emphasizing in his report not only a Franco-Italian Mediterranean kinship but also those elements of the Italian pavilion that he felt resonated with his ideals of a French "third way."

Fig. 2.4 The German Pavilion. Album officiel: Exposition internationale des Arts et des techniques appliqués à la vie moderne—Paris 1937

As France was trying to shape its own modernist destiny at the Paris fair, Italy was seeking to redefine modernism with an Italian flavor. The tension between modernity and tradition, which would become a leitmotif of the planning for E42, played out in both the architecture and the interior displays of the Italian pavilion, particularly in the mix of art, decorative art, and industry.

Officially opened in May 1937, Italy's pavilion involved the purported collaboration of architectural rivals. Marcello Piacentini, Italy's most prominent conservative modernist who led the architectural planning for E42, designed the exterior structure, while Giuseppe Pagano, who represented the opposing rationalist avant-garde wing of Italian architecture, managed the interior design and exhibitions. "In the style of the architect Piacentini," Labbé observed about the pavilion, "the great Roman tradition marries in utmost harmony the geometries of modernism."[43] Situated along the left bank of the Seine, the Italian pavilion comprised a geometric central tower, rising to a height of forty-two meters above the Seine, and a companion horizontal pavilion. These major constituent structures were tied together by two galleries. The building adopted (or, as some alleged, plagiarized) the geometric proportions and look of a toned-down version of the Casa del Fascio (1933–1936) in Como, designed by Giuseppe Terragni, the leading rationalist architect and cofounder of Gruppo 7.[44] "All of the architecture of the Pavilion, and particularly that of the tower," Labbé continued, "was simple and very modern, but with a clear Mediterranean and Italian inspiration. The dimensions were that of a solid classicism."[45] The tower was decorated with four orders of porticos, each surmounted by statues three meters tall, executed by twenty-four sculptors.

The most imposing (and predictable) exterior ornament was a monumental equestrian statue in honor of Il Duce, the *Génie du Fascisme*, cast by sculptor Georges Gori. The internal exhibits were an advertisement for the varied accomplishments of the Duce's Fascist regime. On the ground floor of the central tower was a hall devoted to "Italy abroad," as it was euphemistically titled, designed by Mario Sironi, a leading artist of the Novecento group. It featured a display of Roman monuments in Africa, juxtaposed with the story of Mussolini's conquest of Ethiopia, showing the "economic, artistic, and moral" benefits bestowed by the new Italian

Fig. 2.5 The Italian Pavilion 1937. Album officiel: Exposition internationale des Arts et des techniques appliqués à la vie moderne—Paris 1937

empire. Photographic exhibits on the first floor of the tower advertised public works under Mussolini, especially the "prodigious resurrection" of the once-miasmic Pontine Marshes south of Rome and the creation in their place of "happy" new towns such as Pontinia, Littoria, Sabaudia, and Aprilia.[46]

One of the major attractions of the Italian pavilion was Sironi's immense mosaic in the tower's Hall of Honor, titled *Fascist Work*, in front of which stood a double-life-size bronze *Winged Victory*. As richly described by Barbara McCloskey, the mosaic centers on a large-scale female figure clad in a classical tunic personifying Italy.[47] Surrounding her are scenes depicting the ties between ancient Rome and Fascist Italy, with Mother Italy embracing all. As McCloskey notes, the scenes "feature symbols of the Roman Empire (including the Roman eagle and a classical column), images of work, family life, and the military, as well as mythological figures representing the ancient, Roman, and Christian periods of Italian history." Also highlighted is an image of labor personified by Mussolini as the ideal worker, raising his spade in salute. A halo of light appears around Mussolini's head. McCloskey points out, however, that despite the homage to Fascism and Il Duce, Sironi came under attack from conservative Fascist critics for his controversial expressionistic distortion of human figures, considered to be more German than authentically Italian—a criticism reminiscent of the Nazi notion of "degenerate art."

One of the largest halls of the Italian pavilion, conceived by Pagano, was devoted to an exhibition of industry. Interestingly, Labbé's report mentions the hall in passing and otherwise ignores it. Instead, the official French report on the Italian pavilion focuses on the nonmechanical and nonindustrial displays, such as gardens, food, agriculture, light industry, artisanal work, regional culture, painting, sculpture, and crafts. This emphasis accorded with Labbé's intention that the Paris fair would restore the "soul" and *humanité* in modernity. Expressing a similar desire in Fascism to unify the arts and technology, and to soften the harder scientific and technological edges of modernity (as we will see in E42), the Italian pavilion indeed had much to offer along this line.

Labbé's report marvels at the ability of the pavilion to make one feel "transported at once both to the splendors of the Aventine Hill . . . and to the fields of Ostia, when one reposes in the flowery patios among fat earthenware jars and cascades of clear water, evoking Horace and Virgil."[48] Among such experiences was the pavilion's re-creation of a *fiaschetteria*, or wine tavern, where visitors experienced the sublime pleasures of the Italian countryside and sampled the marvelous range of Italian wines. The pavilion's restaurant offered distinctive Italian cuisine and the fruits and vegetables of the country's various regions, thus supporting the regional theme of the Paris fair. The nearby exhibits of the Ministries of the Press and Propaganda as well as the Gallery of Tourism emphasized "Italian regions, so different from each other, but welded together by a prodigious unity and stunning continuity" under the rule of the Fascist state.[49]

Pleasantly described in Labbé's report is an expansive interior garden, with four great fountains, a powerful evocation of the Mediterranean. Mention is also made of a nearby portico whose walls were covered with murals in praise of the Fascist regime by the Jewish artist and Fascist supporter Corrado Cagli. Tellingly, Labbé's account fails to mention that, with the rising anti-Semitism of the Fascist regime in this period, Cagli's murals were destroyed on orders of the Italian foreign minister.[50] In sum, while reinforcing much of the French government's ideology underlying the Paris fair, Fascist planners of the Italian pavilion were looking quietly

む望を口入手右りよ隅南東　館本日

Fig. 2.6 Exterior View of the Japan Pavilion 1937. Source: Pari Bankoku Hakurankai Kyōkai, 1937 *'Kindai Seikatsu ni okeru Bijutsu to Kōgei' Pari Bankoku Hakurankai Kyōkai jimu hōkoku* [Administrative report of the association for participation in the 1937 'Arts and Crafts in Modern Life' Paris international exposition] (Tokyo: The Association, 1939), plate 1. Collection: Morris Low.

ahead to Mussolini's great World's Fair of Rome in 1942 to announce to the world a distinctive vision of Fascist modernity.

Imperial Japan faced a double task: to demonstrate its mastery of Western modernity without appearing subservient to it. Its solution was to combine tradition and modernity by providing a strikingly modernist, Corbusierian building within which traditional Japanese arts and crafts such as baskets, lacquerware, metalwork, and ceramics were featured. The architect was Sakakura Junzō, who had just spent five years in Le Corbusier's atelier and only recently returned to Japan. Though in his review of the exposition's architecture Laszlo Moholy-Nagy described the Japanese pavilion as "intimately Japanese" despite "its palapable modernity," the traditionalist exhibit within was in sharp contrast to its prizewinning container.[51]

In the July–August 1937 issue of the official magazine *Exposition Paris 1937* the commissioner general for Japan, Suganami Sohji, commented on the challenge of organizing the Japanese exhibit long-distance from Tokyo and the difficulty of painting a portrait of contemporary Japan in miniature. He also lamented the difficulty of capturing a snapshot of a fast-changing Japan. Indeed, war had just broken out with China and greater changes were to come. Suganami pointed out that the Japanese exhibit bore little resemblance to displays at other world's fairs that Japan had participated in. Previously, the tendency had been for Japan to focus on tradition. In contrast, at the Paris Exposition, Japan would display its most modern aspects. An effort was made to show that Japan had opened its doors to Western influence, not only in art but also in business and industry. Suganami wrote that Japan's "greatest ambition is to work in close collaboration with the Occident whose civilization she has assimilated and adapted to her own national genius till she can now be proud of her achievements in the field of modern progress. It will be possible to verify the truth of that statement in 1940 at the Tokio World's Fair where I hope France and the French Press will be largely represented."[52] Suganami saw the Paris Exposition as an opportunity not only to compare the arts and crafts of all participating nations but also to potentially promote new economic collaboration. The brief article was accompanied by a photograph of the exterior of the highly modern Japanese pavilion and a shot of a photomural that graced the interior, showing a montage of touristic images of Mount Fuji and kimono-clad Japanese women. The Japanese display was, for these reasons and others, highly important for Japan, a nation that was preparing to mount its own international exposition at a time when the world was increasingly on a war footing.

As Suganami was also careful to state in the Paris Exposition's magazine, the Japanese pavilion was partly about making cultural statements but it was also intended to promote Japanese business and industry. Some of the Japanese exhibits were displayed in Japan before being sent to Paris. They attracted sharp criticism that was reported in the press. Few of the arts and crafts seemed to showcase scientific innovations; rather, they seemed to draw on long-standing traditional techniques and materials. Indeed, Suganami did not mention science once in his statement. There was thus a disjuncture between the progressive face of Japan as seen in Sakakura's pavilion and the contents that had been chosen by the Ministry of Commerce and Industry. Even the photomontages inside highlighted touristic images of traditional Japan. Architects had decided on the ultimate form of the pavilion but a separate body had taken responsibility for the exhibits.[53]

Compared to the robust ideological presentations of the European powers and even the colorful but bifurcated Japanese exhibition, the US pavilion seemed to many at the time to be disappointing and weak. The Congress of the United States of America was in contemporary terms a day late and a dollar short. The appropriation for the US pavilion came somewhat late for planning purposes, and those who were charged with mounting the exhibition viewed the sum as small. The architect of the building, Paul Lester Wiener, had sixty-eight days in which to work. Wiener was German-born and educated there, in Austria, and in France. He became a US citizen in 1919. For Wiener, the American contribution to modernism was embodied in the skyscraper. To build a real skyscraper for the exposition was impossible. Wiener's goal for the *exposition universelle* was to provide some sense of the skyscraper by capturing the soaring vertical line. "My design," he said, "was to capture this essential quality in its simplest and most dramatic form. The two chief elements which I wished to 'marry' within the United States Building were (1) a symbolic representation of the skyscraper and (2) the dramatic style demanded by exposition architecture."[54] The resulting building was 150 feet tall, from the river to the top of the tower, and was particularly spectacular at night, when the interior lights showed it as a tall, luminous unit. The success of Wiener's efforts was debated then and is controversial now. Edmond Labbé's final report politely lauded the pavilion's "clean forms" that with an "evocative simplicity" showed the American people as "practical and young," but could not resist describing "the broad veranda from which Babbitt could contemplate the comings and goings of cosmopolitan visitors."[55] The journalism professor David Littlejohn regarded the pavilion as "a gigantic cinema of Sunset Boulevard or a hotel on the strip of Las Vegas."[56]

The exhibits inside were almost universally considered less than successful. Littlejohn remarked that "the interior was more funerary than festive." While European visitors may have hoped to see a Main Street soda fountain, they were greeted by an exhibit of California wines that were not available for either tasting or purchase. Indeed, no American goods were available to purchase or to order. Facsimile copies of the Dec-

Fig. 2.7 The US Pavilion 1937. Album officiel: Exposition internationale des Arts et des techniques appliqués à la vie moderne

Fig. 2.8 Eiffel Tower between Soviet and German Pavilions.

laration of Independence and the Constitution were handed out. Exhibits were mainly about New Deal projects like the Tennessee Valley Authority and Boulder Dam. One small exhibit attracted thousands of visitors: a day-to-night model of the forthcoming 1939 New York World's Fair, "Building the World of Tomorrow."[57]

In 1937, the French government's goals were to heal internal divisions by opening opportunity to all classes of society, and to maintain what increasingly looked to be a fragile peace. As a popular postcard from the *Exposition* seems to suggest, France imagined itself as a rational middle way between contending ideologies, the new and powerful forces threatening the stability of Europe. With hindsight, we can judge those hopes barren and forlorn. To some, the image seemed to suggest a France threatened on both sides by newer dynamic forces.

THREE

FANTASIES OF CONSUMPTION AT SCHAFFENDES VOLK

National Socialism and the Four-Year Plan

IN the watershed years between 1936 and 1938, the National Socialist regime set about to reorganize the German economy to facilitate its preparations for total war. With the implementation of the Four-Year Plan, announced by Hitler at the Party rallies in Nuremberg in 1936, raw materials and currency were redirected toward armament production. Hitler promised that German ingenuity and technological innovation would provide the jobs and products needed to lift the standard of living for the Volksgemeinschaft and would liberate Germany from economic dependence on imported foreign goods and natural resources.[1] Nazi ideology defined German economic and scientific activity in racial and spiritual terms, and thus starkly contrasted it to what its anti-Semitic vitriol denigrated as the rootlessness and "decadence" of Jewish and American consumer capitalism.[2] Yet as Germany's employment numbers rose as a result of its armament-driven recovery, its citizens wanted once again to partake in the kind of vibrant consumer marketplace they had enjoyed before the 1929 economic crash. The Nazi leadership balanced consumer desire with the economic necessities of its extensive military preparations by successfully harnessing the politics of consumption to a powerful propaganda campaign of racial belonging and civic duty. The German Volk had to be conditioned to make consumer choices according to the collective benefit and priorities of the nation and race, rather than succumbing to individual desire or need. By promising a future prosperity (which was never actually realized), the Nazi leadership sold the German people on the paradox of enticement and sacrifice simultaneously.

One of the most significant ways that the Nazi regime set about to educate and encourage citizen-consumer collaboration was through the staging of large industry and culture exhibitions. These fairs provided various forms of window-shopping and entertainment, while exalting industrial production, technology, and art as the direct spiritual expression of a unified Volk revitalized by National Socialism. During the 1930s the number of exhibitions and trade fairs dedicated to art and industry in Germany skyrocketed, as both private companies and the National Socialist regime orchestrated manifestations of patriotic production and consumption. Foregrounding Nazism's supposed "liberation" of technology and science from parasitic capitalism, these fairs sought to align the German citizen with the priorities and values of the fascist state. This chapter will focus on one of the largest and most significant of these fairs, the 1937 "international" exhibit Schaffendes Volk in Düsseldorf, which according to official records attracted almost seven million visitors.[3] The fair's title was translated at the time as "A People at Work," though this phrase inadequately conveys the meaning of the German verb schaffen, which points more to the qualities of being productive, creative, and industrious. Schaffendes Volk celebrated advanced technological production and the spirit of German labor, and announced the rebirth of the German economy under Nazism. The pavilions celebrated Germany's efforts to achieve self-sufficiency through technological innovation and highlighted the development of synthetic replacements for raw materials. It exalted the importance of the male laborer and the female homemaker and encouraged the German population to both consume and conserve in ways that aligned

Fantasies of Consumption at Schaffendes Volk 31

Fig. 3.1 Schaffendes Volk exhibition brochure, Stadtarchiv Düsseldorf, NL Emundts 1854.

with the goals of the state. Germans were told that they could experience shopping and leisure under the Third Reich as normative pleasures, but would also be called upon to renounce previous habits and replace certain products for the long-term public good.

Schaffendes Volk opened in May 1937, the same month that the Paris Exposition Internationale officially opened. Despite claims by the Nazi organizers that Schaffendes Volk was not organized to compete with the Paris fair, it is clear that the timing was deliberate.[4] As the *New York Times* reported before the opening, "a vast German fair was being built in secret" as a direct rival to the "oft-delayed" Paris exposition (though Schaffendes Volk, like the Paris Expo, was also not completed at the time of its official opening, a fact not discussed in the press). The writer noted that with over forty exhibition halls, thirty pavilions, twenty restaurants and cafes, and a huge amusement park, it was the biggest exposition ever staged in Germany. Covering as much acreage as the Paris fair, its grounds stretched more than a mile along the bank of the Rhine.[5] The timing of the fair established a symbolic competition with France, but it was also hoped that the timing would encourage international visitors bound for the Paris fair to make the extra trip just over the border to visit Schaffendes Volk. For this reason, the German exhibition and convenient travel to Düsseldorf via travel services such as Thomas Cook or the German railway were advertised in the pages of the Paris exposition's publications, in French magazines and newspapers, and in posters hung around the city. In order to guarantee the success of Schaffendes Volk, the German Werberat, which controlled exhibition and trade fairs, banned all other competing industrial exhibitions from taking place in Germany between 1935 and 1939, including an exposition that had already been planned for Dortmund in 1937.[6]

Schaffendes Volk was an "international" exhibition in name only: other foreign nations did not have their own dedicated pavilions or displays, as was the case of the 1937 exposition in Paris or the 1939 fair in New York.[7] Schaffendes Volk, however, was clearly considered in the world fair category. The Nazi organizers stressed that the commissioners of other recent and upcoming world expositions—Brussels, Chicago, New York, and Rome—upon visiting Schaffendes Volk found it exemplary and believed it offered a new way forward for exhibition design. Schaffendes Volk was

Fig. 3.2 "Prince Chichibu visits the Schaffendes Volk exhibition," *Reichsausstellung Schaffendes Volk Düsseldorf 1937. Ein Bericht.* Edited by Dr. E.W. Maiwald. Volume 1, Düsseldorf, 1939.

estimated to have attracted between 250,000 and 300,000 foreign tourists and business people, as well as over three hundred foreign journalists. According to the fair's final report, over eight thousand guided tours of the fairgrounds were offered to visitors in twenty-two different languages.[8]

Among the notable guests at the opening of Schaffendes Volk were ambassadors and diplomats from France, Holland, Belgium, Yugoslavia, and Argentina. During the course of the exhibition, there were also individual visits by prominent dignitaries, including Prince Chichibu, the brother of the emperor of Japan, and the duke of Windsor (who had recently abdicated the position of king of England to marry the American socialite Wallis Simpson). Both men had strong pro-Nazi sympathies.[9] These visits were celebrated in the German press, and photographs of the distinguished guests admiring various industrial exhibits were featured in the official illustrated book about Schaffendes Volk published by its commissioner E. W. Maiwald.[10]

For the purposes of this chapter, I will discuss Schaffendes Volk through a few lenses: in terms of the goals stated in Hitler's Four-Year Plan, in light of the instrumental role consumer culture played in manufacturing political consent, and to argue for the importance of reasserting historically based notions of modernist aesthetics and ideology and their complex intersection with the reactionary policies and ideologies of the fascist state. Schaffendes Volk is an excellent heuristic device by which to examine the changing interface between the public and private spheres under National Socialism, whether considering government regulation and private business interests, political propaganda and consumer behavior, or the translation of ideological goals into visual design. Comparison to the German participation at the Paris World's Fair is also instructive for understanding the ideological scope of Schaffendes Volk. The German pavilion that Albert Speer built for the Paris fair was the largest manifestation of German industry and culture outside the borders of the Third Reich, and its visual iconography and industrial displays communicated that the kind of nation-building actively being constructed at Schaffendes Volk had in fact already been accomplished. As a result of its positive reception in Paris, the Third Reich gained significant political cachet, which it then translated into economic, diplomatic, and eventually military gains.

Schaffendes Volk, the Werkbund, and Reactionary Modernism

The idea for an art and industry exhibition in Düsseldorf was initially conceived by the German Werkbund before the ascent of Hitler in 1933. The Werkbund intended the exhibition as a showcase for German design, a celebration of the partnership between product manufacturers and design professionals, and the integration of traditional craft techniques with industrial mass production. In 1935 the Nazi Party took control of the exhibition, placing it under the patronage of Reichsminister Hermann Göring, whom Hitler had also put in charge of the Four-Year Plan. Göring became the central figure in economic matters for the Third Reich, increasing the already extensive reach of his power and further tying Germany's new economy to massive rearmament. It was at this point that the exhibition took on the name Schaffendes Volk—the result of a public competition— and shifted its emphasis to the potential of technology and to redirecting consumer behavior toward Hitler's economic goals. Germany was to achieve autarky by increasing agricultural production, retraining key labor sectors, controlling wages and prices, slashing imports, and reducing dependence on foreign raw materials through the development of synthetic replacements. The organizers of Schaffendes Volk were given only seven months to transform the exhibition from a *Werkbund Siedlungsschau* (housing development exhibition) into a showpiece for the regime's ambitious economic and political agenda.

Hitler's declared goal of self-sufficiency did not mean he was relinquishing Germany's colonial demands. As the exhibition would illustrate, territorial expansion was central to Germany's ideological and economic goals. By the time the Nazi Party took control of Schaffendes Volk, the Werkbund itself had already undergone the process of *Gleichhaltung*, first through its placement under the Kampfbund für deutsche Kultur (Combat League for German Culture) and then under the Reichskulturkammer (Reich Chamber of Fine Arts). The "Nazi leadership quickly saw the potential of modern commodity styling . . . in manufacturing winning cultural self-images of achievement and prosperity."[11] Reciprocally, National Socialism appealed to the neo-corporatist aspirations of many within the Werkbund. As control of Schaffendes Volk was transferred to the party leadership, significant aspects of Werkbund culture were adopted to suit the regime's own political and economic ends. Nazi cultural policymakers made use of the Werkbund's language of social reform and its idealization of functionalism, as well as the concept of *Arbeitsfreude*, or "joy of work." The regime continued the Werkbund promotion of German craftsmanship at home and abroad and its commitment to training workers in various craft trades. The collaboration between Werkbund members and the Nazi office of Beauty of Labor also maintained an emphasis on hygiene and recreational facilities for the Volk.[12] At Schaffendes Volk, this continuity was evident in the numerous exhibits of industrial design products that exhibited a distinctly modern aesthetic associated with Werkbund and Bauhaus production—steel tube chairs, simple tableware, geometric-inspired lighting, and other home furnishings. Leading German firms had already established worldwide reputations and export markets in this area, and "the Nazis in no way wished to jeopardize such a lucrative source of profit and international good will."[13]

Another reason modern design was tolerated under National Socialism

Fig. 3.3 Exhibition in the Steel and Iron pavilion (Gemeinschaftsausstellung der Deutschen Eisen- und Stahlindustrie), photo collection, Stadtarchiv Düsseldorf.

was that domestic consumers still desired these products. As Despina Stratigakos has noted, the younger generation of Germans had grown up with modern design, particularly in urban areas where it had gained widespread popularity. This demographic group "did not want homes that looked like those of their grandparents." [14] Younger consumers were prime targets for Nazi propaganda because they were getting ready to set up house for the first time on their own. The German government wanted to encourage marriage and reproduction while reducing unemployment by offering these young couples interest-free loans if the bride agreed to give up her job and stay home to have children (further incentives followed upon the births of subsequent offspring). The loans, worth up to seven months' wages for an average worker, were offered in the form of vouchers that could be used to purchase furniture and other household items. [15]

From an architectural standpoint, the main exhibition halls also reflected modernist typologies and construction methods. The modernist aesthetic further supported the fair's celebration of German progress, productivity, and future prosperity. Many of the exhibition buildings had large expanses of glass windows, steel infrastructure, and flat rooflines;

Fig. 3.4 Fritz Faulenbach, exhibition hall for the Cement Association of Düsseldorf. (Hüttenzement-Verband GmbH, Düsseldorf). *Reichsausstellung Schaffendes Volk Düsseldorf, 1937* p. 166.

these included the hall dedicated to architectural construction methods designed by Klaus Reese and the pavilion built for the Mannesmann Company by the modernist architect Emil Fahrenkamp (known for his iconic Shell-Haus built in 1930 in Berlin). There were also numerous hangar-like structures, such as the "Stahl und Eisen" (steel and iron) pavilion, that provided immense warehouse spaces for the display of large machines, train cars, and steel foundry equipment. As reported in the *Industrieblatt Stuttgart*, the dimensions of the pavilion (an open span of eighty-five meters; fourteen meters tall at its highest point), demonstrated the potential of steel and iron construction.[16] Certain pavilions made deliberately bold modernist architectural statements, most notably the building designed by Fritz Faulenbach for the Cement Association of Düsseldorf. Its streamlined cantilevered roof, which rose ten meters up from the ground and measured eighteen meters across, demonstrated the versatility of cement as an architectural medium. The freestanding reinforced concrete plate was touted in the press as "the first of its kind in the world to be built on such a scale."[17]

The interior design of most of the exhibition halls was also intended to highlight the "cutting-edge" nature of the displayed goods, as was the case

Fig. 3.5 Interior of the Mannesmann pavilion. Photo collection, Stadtarchiv Düsseldorf.

in previous German industrial exhibits, most notably the 1934 Deutsches Volk–Deutsche Arbeit, which counted among its collaborators Mies van der Rohe and Walter Gropius. Machines were showcased as modern aesthetic objects in their own right: factory equipment and technical products of major companies such as Borsig, Mannesmann, and Demag were either mounted on pedestals like sculpture or formed into shapes resembling spiraling constructivist sculptures or maquettes for imagined skyscrapers. Photographs of Schaffendes Volk taken by Albert Renger-Patzsch and Hugo Schmölz, who were previously associated with Neue Sachlichkeit (new objectivity), completed the conversion of these machined objects into pleasing aestheticized forms (Schmölz's images were published in the architecture journal *Moderne Bauformen*).[18] Several of the industrial designs and exhibition displays shown in Düsseldorf were duplicated by German companies at the 1937 Paris Exposition. These objects were similarly celebrated for their modernist appeal in full-page, black-and-white photographs by the French photographer Hugo Herdeg, whose images appeared

Fig. 3.6 Display showing possible applications for synthetics. Photo collection, Stadtarchiv Düsseldorf.

in the vanguard art magazine *Cahiers d'Art*. The aestheticization of German technology in this journal, well known for championing the work of modern artists and art movements, speaks to the ways the slick surfaces and forms of these machines were dramatized through exhibition design and photography. Lost in the aestheticizing process were vital political and economic subtexts: the fact that many of these industrial products and the companies that manufactured them were of central importance to German rearmament efforts and preparations for war.[19]

Modern advertising and exhibition design strategies were exploited at Schaffendes Volk despite the regime's tight regulation of the media and its intolerance for modernism in the fine arts. Nazi ideologues understood what powerful propaganda could be produced by the paradoxical combination of romanticized anticapitalist rhetoric with the technological optimism of the European vanguard. The organizers claimed to "break new ground in the exhibition industry," because of their use of oversized photographs, photomontage, dioramas, and three-dimensional models and placards jutting out into the viewer's space from various angles and perspectives.[20] According to Commissioner Maiwald, Schaffendes Volk was intended to herald not only "the beginning of a new economic era for Germany, but also the start of a new era in exhibition design."[21] These exhibition practices show how "reactionary modernism" could be successfully implemented within such a utilitarian context.[22] What the German press referred to as "audience-grabbing" techniques were used to promote goods from synthetic materials and fashion to electronics and laundry detergent, as well as political slogans and party doctrine. The medium of film was also used in various sections of the fair for propagandistic effect. For example, in the exhibition organized by the Reich Education Ministry (Reichserziehungsministerium), a large oval screening room was built in the center of the gallery and its outer walls covered with photomontages and other didactic materials.

In his final report to the German press Maiwald contrasted what he called the innovative and forward-thinking exhibition practices of Schaffendes Volk with the outdated model of the world's fair taking place in Paris. He asserted that "Düsseldorf certainly had no ambition to compete with the Paris exposition," but he denigrated the French competition nonetheless, arguing that it "had nothing new to say about the pace of technological development." The Paris Exposition, he declared, "was not a trendsetter in any way" but rather the "last example of a bygone epoch in exhibition making." The Germans' innovative approach to Schaffendes Volk, on the other hand, established a new direction and standard for future world exhibitions. Instead of "an inchoate jumble of caricatured architectural styles and unrelated themes," the world's fairs of tomorrow must represent in a systematized manner "the progress, future challenges and opportunities of each country, so that they can be realistically depicted and compared."[23] What Maiwald's remarks fail to address is the retrograde design of the German pavilion at the Paris Exposition, for the Deutsches Haus was hardly future-looking. As I have argued elsewhere, the pavilion's architecture and interior décor rejected modernist aesthetics, appealing instead to conservative bourgeois taste.[24] When visitors stepped through Speer's stripped-down neoclassical façade, they were met by an ostentatious interior reminiscent of an oversized nineteenth-century parlor. The interior was decorated with immense mosaics and stained glass to evoke both classical and medieval precedents. The swastika-patterned wallpaper, ornate chandeliers, oil paintings, and old-fashioned wood vitrines created an incongruous environment for the many modern technological products displayed alongside the fine china, leather goods, and other luxury items for sale by German manufacturers. Why such a difference for two exhibitions on culture and technology opening simultaneously? Maiwald was involved as a commissioner in the organization of both exhibitions, but the two projects had very different political agendas. The pavilion in Paris was directed at foreign audiences in an effort to convey Germany's stoicism and dignity, and to instill a distant sense of awe and wonder. It was also intended to reassure visitors in Paris of Germany's peaceful intentions and commitment to rapprochement. Schaffendes Volk, on the other hand, was intent on civic engagement and interpellating German citizens into the national mind-set required by the Four-Year Plan.

It should be underscored that the Nazi desire to transform German society through racial persecution and the implementation of new economic and political structures also constituted a form of modernism. Under National Socialism, reactionary modernism went beyond the aesthetic appearances in cultural production to its core ideological underpinnings of social transformation and regeneration. In the spirit of an avant-garde, Nazi ideologues and technocrats promised to bring order and spiritual renewal to the chaos and rootlessness of life under capitalism. As Peter Fritzsche has noted, "Although the Hitler regime cannot be adequately described as merely a German version of Beveridge's England or Roosevelt's America, the Nazis operated in the subjunctive tense, experimenting, reordering, reconstructing, and it is this spirit of renovation that qualifies National Socialism as modern."[25]

The Four-Year Plan, the Colonial Question, and the "Necessity" of Lebensraum

When visitors entered the fairgrounds of Schaffendes Volk, a long procession of flags flanking a ceremonial plaza directed them toward the Honorary Hall of Working Volk (Ehrenhalle des werktätigen Volkes). This exhibit was intended as an introduction and ideological framing for the fair as a whole, where the regime promoted its vision of Germany's future, free from any reliance on foreign products and imported natural resources. Quoting Hitler's speech on the Four-Year Plan, the exhibition press office announced: "This is a turning point in the history of the German Volk. The economic dependency of Germany on international suppliers of raw materials and the irrationality of the commodities market is a constant threat to our political freedom and is unbearable for a healthy emerging nation. Our people must be free."[26] The Four-Year Plan was being promoted at the same time in Berlin at the propaganda exhibition *Gebt mir vier Jahre Zeit!* (Give me four years!), which had opened the previous month.[27] The German people were promised that if they committed to hard work and dedication, their country would finally be released from the whims of international trade. German labor, working alongside German science, had the power to transform the simplest of raw materials available in their native land into a spectrum of "wonder products." The most important of these were synthetic alternatives for natural rubber, fuel, and textiles.

The goal of the Four-Year Plan was to halve Germany's import bill as quickly as possible. The regime would then use the resulting reserve of foreign currency toward the purchase of raw materials required for the manufacturing of military equipment to build up the country's war machine.[28] This plan meant that Germans would have to deprive themselves of desired imports, but also of some domestic goods and foodstuffs that were needed as exports to generate revenue as foreign credits. The Third Reich initiated a partial austerity program in both consumer goods and food staples, and pushed for a recycling program (as in reclaiming scrap metal). Although the underlying purpose of the Four-Year Plan was to ready the nation for war by directing the entire economy toward the production of armaments, Schaffendes Volk cast the new plan in entirely different terms that highlighted consumer benefits. Rather than emphasizing the hardships to come, it promised a new era of consumption based on the most advanced technological research and achievement, promising riches out of the most modest and common materials indigenous to Germany. As Commissioner Maiwald would proclaim in his final report, the success of Schaffendes Volk was the conversion of "6.9 million exhibition visitors into 6.9 million propagandists for the Four-Year Plan."[29]

The newly imagined relationship between the German Volk and German technology was given visible form in the Ehrenhalle. The exhibition design centered on the installation of four massive pillars, each filled with one of the major raw materials found in Germany: coal, wood, iron, and stone and earth. This section, called "Wunder der Technik," introduced visitors to the miraculous potential of products that could be manufactured from these basic elements through a series of immense diagrams, flowcharts, and displays. The Association of German Engineers (Verein Deutscher Ingenieure), worked with architect Gerhard Graubner and artists M. Janssen and W. Geißler to design the Ehrenhalle, in cooperation with various industrial organizations. A detailed survey of the exhibition was published in the association's journal, which celebrated the exciting social and economic potential of Germany's new technological "revolution." It explained how something as simple as one kilogram of iron ore pumped from the earth and worth only a few pennies would increase in value as it moved up the ladder of processing and manufacturing, finally "surpassing the price of gold" once formed into desirable products or machinery.[30] The austerity measures put in place by the Third Reich, its restrictions and rationing of raw materials, were framed as the actual source for future ingenuity, necessity spurring on an even more revolutionary and creative innovation. In his speech at the opening of the exhibition, Göring likewise announced that the Four-Year Plan was the "beginning of a new technological era" for the country.[31] Not surprisingly, Schaffendes Volk made it appear as if all the technology on display had been invented or discovered only after the National Socialists had come to power.[32]

The organizers of the Ehrenhalle used a range of media in orchestrating the exhibit—photomontage, graphics, and instructive panels—to convey the interdependence of the twin powers of German technology and

Fig. 3.7 "Labor Creates Value from Coal." Didactic display in the Ehrenhalle. Photo collection, Stadtarchiv Düsseldorf.

the German workforce. German labor was idealized for its dedicated work ethos with such slogans as "Dignity and Grandeur" and the "Beauty of Labor." The propaganda narrative followed that after the defeat of World War I and the humiliation of reparations, the German dedication to work under National Socialism had once again restored confidence to the Volk. Surrounding the requisite busts of Hitler and Göring, immense frescoes brought together idealized images of scientists, miners, steelworkers, engineers, and farmers working side by side in a corporatist utopia. In the

adjacent exhibition devoted to "Iron and Steel," the large wall text reinforced the heroism of German labor with the declaration "Iron Workers
Will Secure Our Freedom and Honor." The Schaffendes Volk catalogue
referred to this pavilion as "a symphony of the Steel Industry," a motif
echoed repeatedly in the press, which headlined the exhibition as a whole as
a "Symphony of Work."[33] The imagery was employed as well in the pavilions built by individual companies such as the Deutsche Maschinenfabrik
in Duisburg, a producer of heavy machinery (known by its abbreviation
"Demag"), whose façade featured a fresco by artist Fritz Buchholz exalting
the noble worker and engineer. Four immense figures representing various
kinds of labor stood at a height about five times that of an average adult
visitor. The fresco functioned as an allegorical conceit, with each of the
men holding an attribute to his trade in shallow relief against a neutral
background. The two manual laborers who hold a jackhammer and anvil
are shirtless, their musculature exaggerated by the graphic simplification
of their forms. Yet the figures stand shoulder to shoulder as equals with
the engineer, who is fully clothed and holds a sheet of blueprint plans. The
two pairs turn toward the center of the façade where a massive machine
has been installed between the two sets of entrance doors. The metonymic
association of man and machine was common in Nazi iconography, the re-
fining and forging of metal and steel akin to the "purification" and "fortifi-
cation" of the German Volk under National Socialism. Immense allegorical compositions exalting the worker and engineer likewise dominated the
program of the German pavilion at the Paris Exposition Internationale.[34]

The implementation of a shared ideologically driven iconography by
the Nazi party and German private industrial concerns such as Demag at-
tests to how closely big business collaborated with the regime. The central
role of IG Farben in the development and production of synthetics, for example, demonstrates how much private corporations both contributed to
the industrial war effort and benefited from the exigencies of the Four-Year
Plan. As noted by the second-in-command at IG Farben, Dr. Georg von
Schnitzler, the possibilities opened up by the Four-Year Plan "undoubt-
edly exercised a great fascination upon our technical people." The huge
amount of resources funneled to IG Farben by the Third Reich meant that
potential ideas that wouldn't have had "any practical realization as long
as a normal economy existed" all of a sudden "became realizable, and the

Fig. 3.8 Façade of the Demag pavilion with frescoes by F. Buchholz. Photo collection, Stadtarchiv Düsseldorf.

most fascinating prospects seemed to lie ahead."[35] Key among the synthetics celebrated at Schaffendes Volk was the display of IG Farben's Buna, or synthetic rubber, which was manufactured from coal and limestone. Described by the Schaffendes Volk's press office as a "wonder of German ingenuity," it noted that the exhibition didn't just present boring statistics about its manufacturing but brought the excitement and future potential of Buna alive for the visitor with the display of products purportedly ready for implementation.[36] (IG Farben would of course become known most infamously for the production of Zyklon B gas, which was used to kill millions of Jews and other victims in extermination camps.)

Germany's desire to achieve economic self-sufficiency with the Four-Year Plan was intimately linked to revived claims over its former colonial territories. The production of synthetics was one way to reduce Germany's dependence on foreign imports; harvesting natural resources from colonial holdings was another. One week after the official opening of Schaffendes Volk, the Colonial Exhibition opened on the grounds of the Düsseldorf zoo, organized to coincide with an important meeting of the Reichskolonialbund (Reich Colonial League). The Reichskolonialbund also organized an exhibition hall on the main grounds of Schaffendes Volk. The symbolic placement of their main exhibition at a zoo for animals demonstrates, not surprisingly, that the dehumanizing and racist logic of colonialism thrived under the Third Reich. Speaking at the exhibition's inauguration, Admiral Rüman, chief executive of the Reichskolonialbund, underscored the connection between the goals of Schaffendes Volk and Germany's "colonial question." In his speech, Rüman reiterated Hitler's call for the return of Germany's colonies, declaring that the Third Reich was still the "rightful owner" of these territories since they had been wrongfully expropriated by the terms of the Versailles Treaty. He explained that both the Colonial Exposition and Schaffendes Volk demonstrated how important colonial possessions were in "solving the commodity issue." The two exhibitions illustrated the importance of "supplying raw materials to the industries of the motherland" and "the great benefits in all areas the colonies had offered Germany in the past."[37]

The necessity of German territorial expansion was further developed in another significant pavilion at Schaffendes Volk, dedicated to the subject of "Deutscher Lebensraum," or "German living space." *Lebensraum* was the ideological term used to describe the National Socialist intention to "redistribute people and work" by occupying lands primarily to the east where they could establish German settlements. The Deutscher Lebensraum exhibition in many ways provided an overarching and unifying ideological matrix for Schaffendes Volk, one shot through with the mythology of racial superiority. Through the lens of a shared *völkisch* destiny, the visitor could project these larger spiritual and political values onto the displays in other halls, which otherwise might have appeared as mere collections of disparate industrial goods. Exhibited within the Lebensraum Hall were large-scale maquettes for new and planned German settlements, as well as models of ongoing infrastructure projects including the German railway system and the Autobahn. These extensive projects were central to physically uniting and modernizing German territory. The very process of building the Reichsbahn was made synonymous with the racial uniting of the German labor force, as indicated in the use of the racialized term *Volksgenossen* to describe the "700,000 men working for Volk and Fatherland" to construct the railway system.[38]

The use of maps to visualize racial bonding under National Social-

Fig. 3.9 "700,000 *Volksgenossen* working for Volk and Fatherland to construct the railway system," Deutsche Lebensraum pavilion. Stadtarchiv Düsseldorf, NL Ebel 123.

ism permeated the Lebensraum Hall. The centerpiece was a large three-dimensional map of Germany depicting state and county borders, highways, provincial districts, major roads, rivers, lakes, castles, railways, cities, and towns, served as the exhibit's centerpiece. The comprehensive 3D relief formed the template for nine other maps and photomurals that covered a total of four hundred meters of wall space, each map highlighting specific themes and justifications for German territorial expansion. The supposedly unhealthy overpopulation of German cities and the severe shortage of housing were blamed largely on the harsh economic conditions created by the Versailles Treaty. Diagrams and large maps were used to justify the transformation and reorganization of German cities, towns, and transportation and food production networks, and to validate entirely new German settlements in foreign territories. The case for German expansionism was addressed in other exhibition halls as well, including those dedicated to "nutrition, food, and household products." The Lebensraum Hall included an outline of Germany completely filled with illustrations of people and factories. The caption read: "Germany: A Volk without Space." In the adjacent panels infographics indicate how much more crowded Germany was per square hectare than the United States, France, and the Soviet Union.[39] The vivid portrayal of these statistics was complemented by the symbolic placement within the exhibition of a reproduction of Alexander Altdorfer's *The Victory of Alexander the Great (The Battle of Issus)* from 1529. The painting depicts the seemingly endless parade of thousands of soldiers engaged in the slaughter and defeat of the Persian army in 333 BCE. Commissioned originally by the duke of Bavaria, the reproduction's inclusion in the context of Germany's "struggle" for Lebensraum immediately evoked the nobility and necessity for total mobilization.

The promise of Lebensraum and a higher standard of living was given concrete form at Schaffendes Volk through the construction of a vast new housing development, or *Siedlung*, in an area adjacent to the Düsseldorf exhibition grounds (a neighborhood that continues to exist more or less intact today). The new development was considered part of the exhibition itself, with eight of the eighty-four single-family units serving as model homes that visitors could tour. The model houses were equipped with state-of-the-art home technology and decorated in a pastoral völkisch aesthetic. Each house also had a garden plot so a family could grow their own food for optimum self-sufficiency and nutrition. The role of specific German companies in the design of these homes was made clear in the signage and advertising. For example, one unit was outfitted entirely with products manufactured by Henkel—a major German corporation whose pavilion in the main exhibition grounds was one of the largest at the fair. The housing development also included an artist's colony, consisting of a gallery space, three shops, a restaurant, ten artist homes, and twelve studios.[40] Despite the extensive propaganda campaign celebrating the utopian dimensions of the new community, construction of the development fell seriously behind schedule, and when Schaffendes Volk opened, the majority of the houses had not yet been built. The eventual occupancy of the homes also did not live up to its populist rhetoric, with party and city officials grabbing the most desirable properties for themselves.[41]

Engendering Consumer Behavior and the Politics of Marketing Psychology

The technological advances promoted in the model houses were marketed primarily to female customers, as both the regime and private companies realized that they were the ones who largely determined household consumption. Schaffendes Volk was advertised as offering women "countless exciting ideas and inspiration," and they were addressed specifically in the exhibition displays, décor, and wall texts. They were also invited to participate in the "mothers' school" sponsored by the German Women's Bureau (Deutsches Frauenwerk). These classes were advertised as offering women useful "wisdom to keep pace with the way modern industry revolutionized kitchen work and home financial management."[42] In her book on German national identity and domesticity, Nancy Reagin convincingly demonstrates how the Third Reich actively sought to engage women in the goals of the Four-Year Plan. Housewives became the targets of intensive propaganda, as "the Nazi party and state agencies sought to 'spin' the ordinary housewife's experience of the marketplace, redirect her purchasing choices, and increase her work load."[43] The regime recognized that a woman's role as homemaker and consumer could greatly contribute to the autarkic goals of the state through thrifty economizing of their purchases and time-saving adjustments to their housework routines. Though the sphere of consumption was marked by a limited selection of goods, the role of women as consumers was nevertheless framed as empowering because they could make scientifically informed and politically responsible choices. Female comportment then in the private sphere was given symbolic political importance in the public realm. The housekeeping tips and propaganda that supported the Nazi Four-Year Plan were largely distributed through the Women's Bureau, which set up a chain of advice centers, radio programs, and exhibitions through which they distributed recipes and weekly menu plans and pushed consumer education. Among the consumer issues prioritized at Schaffendes Volk were the prevention of food spoilage, the proper use of ersatz products, and the importance of recycling (particularly metal, which was in high demand due to weapons production).[44]

It was the women who had to be convinced that synthetics constituted a worthy purchase, especially the older generation who still remembered how poorly synthetics had performed as substitutes during World War I. The largest section of the fair devoted to synthetics focused on the textile industry, and it was these products that also attracted the most attention abroad in the foreign press. Displays introducing rayon, acetate, and synthetic silk all hailed the positive impact these products would have on the German economy. While the use of synthetic silk in the production of women's garments gained some acceptance during the 1920s, the synthetics being pushed under the Four-Year Plan were intended mostly as substitutes for the staple fabrics of wool and cotton, which previously had been furnished to the German market primarily through imports.[45] At Schaffendes Volk, images of young, carefree men and women dressed in a range of casual, sporty, and elegant attire were juxtaposed with figures covered only by images of foreign currency debt, marked with a large "X" by a trompe l'oeil hand. The accompanying wall text contrasted past and present fashion: "Earlier our clothing made us dependent on speculation, foreign trade and debt. Today, by removing raw materials from clothing produced entirely out of Vistra Rayon, it not only saves us from foreign debt and speculation, but also gives us more beautiful and better clothing. More beautiful and better clothing makes for happier people." In reality, however, serious shortages of all kinds of textiles continued to worsen in Germany despite government pressure to increase the production of synthetics. Furthermore, the Nazi decrees requiring higher percentages of synthetics led to a steep decline in fabric quality. Regulations were introduced that forbade clothing manufacturers from attaching labels with cleaning instructions to garments, so that the inferior quality of the synthetics being pushed by the government would not be exposed. Clothes made from synthetics didn't hold up in the wash, were sensitive to high temperatures, and often leaked color.[46]

The contradictions between the advertising of German synthetics and the reality of living with them were indicative of the complicated con-

Fig. 3.10 Display about the synthetic fabric rayon by IG Farben in the Textiles pavilion. Photo collection, Stadtarchiv Düsseldorf.

sumer lifestyle the Third Reich was trying to sell to its population. The regime recognized that Germans still wanted to participate in consumer culture and that it couldn't politically afford to withdraw key markers of a modern lifestyle from circulation. In walking the line between the desirable goods, fashion trends, and political-economic necessities, Germany's leaders backed products and leisure activities that didn't demand foreign currency for imports or the expenditure of raw materials. In addition, Nazism's so-called split consciousness allowed consumer pleasure to continue under the regime, despite its constant ideological attacks on trade and finance, which it associated with decadent Jewish enterprise, all the while insinuating that sacrifice was not only noble and patriotic but necessary. Employing modern advertising techniques, the Hitler regime used consumer culture to stabilize the country while it asked for patience in delivering on promises of future abundance.

The high visibility of the Coca-Cola Company at Schaffendes Volk exemplifies both this multipronged political strategy and the regime's ideological hypocrisy. The soft drink manufacturer built a model bottling plant near the center of the fairgrounds, which visitors could watch in process through a glass wall. The machinery washed, filled, and capped four thousand bottles an hour, and visitors could line up at a sixteen-meter-long service counter to buy a bottle of the soft drink to sample. The factory demonstration proved to be very popular with the public and was frequently written about in the press. The celebration of Coke at Schaffendes Volk was a mutually beneficial arrangement: it contributed to the appearance of continuity and consumer ease under Nazism, while legitimating Coca-Cola as an acceptable addition to German *Trinkkultur*. Jeff Schutts has noted that consumers "saw the soft drink as an affordable taste of the new, more affluent Germany which they were promised in Nazi propaganda."[47] The participation of the company at Schaffendes Volk marked a significant turning point for the beverage's sales. The half-million bottles sold at Schaffendes Volk were not nearly as important as the amount of positive publicity the company received. "Coke wholesalers across the country were reporting that after visiting the exhibition bar-keepers who had refused Coca-Cola for years were now ready to add the soft drink to their menus." The company declared that Coca-Cola's good name now "radiated throughout Germany, as proven by frequent mention of our company's participation in Düsseldorf by the general public in the further reaches of the Reich." Coca-Cola became thoroughly integrated into Schaffendes Volk's "mirror image of German culture and German commerce."[48] Coke's success does not mean that the beverage somehow found a place in an ideology-free consumer sphere. Rather, the Cola-Cola franchise in Germany managed to completely "indigenize" its product through an aggressive rebranding and public relations campaign that em-bedded the soda in the imaginary of the German *Heimat* and Nazi princi-ples. The Germanization of Coke meant that it was perceived as something "one could enjoy in the service of the Volksgemeinschaft": advertisements stressed that the soft drink helped drivers stay awake and safe at the wheel, and that it refreshed German workers during quick breaks from the job without the moral stigma of tobacco.

The Nazi regime supported the use of mass media to promote com-mercial interests and the promise of the "good life," and the Werberat der Deutschen Wirtschaft (Advertising Council for the German Economy) carefully regulated every ad's copy and image to make sure the content suitably represented Nazi racial ideals.[49] Advertisements were expected to support the notion that producers and consumers were part of the same mythical racial community and therefore shared the same interests and values.[50] Purely mercantile "fairground-style marketing" was denigrated in favor of the noblesse of a well-thought-out and coherent advertising campaign based on a solid brand platform. The opposition between these two kinds of marketing approaches paralleled in many ways the long-standing division between *Zivilisation* and *Kultur, Politik* and *Geist*.[51] Goebbels's version of propaganda and his creation of "Brand Hitler," however, represented an even more "modern edge" in the evolution of political spin by reconciling the rejection of Enlightenment reason with an embrace of contemporary media—the reactionary modernist marriage of the irrational with technology.[52] The manner in which the leaders of the Third Reich manipulated the politics of consumption was far more successful than anything attempted during the Weimar period, despite the fact that the National Socialists' promise of plenitude never materialized. As Michael Geyer maintains, "For most post-war Germans, the Nazi years of peace, employment and victories (1936–42) were the best years of their lives (especially when compared to the Depression before and total war after). Well-being and 'fantasies of unlimited consumption' were linked with sentiments of superiority, an excessive war, and of belonging to a modern nation."[53] The regime therefore attempted to control consumer behavior by promising a utopian future of abundance while simultaneously delaying material gratification for the benefit of its rearmament agenda. Advertisements during the Third Reich paradoxically sold "deprivation" and "enticement" simultaneously.[54] These images privileged collective over individual needs and suppressed representations of class differences that potentially threatened the regime's imperial ambitions. To ensure Germany's survival in an "expanded" Lebensraum, the Third Reich required the unreserved commitment of the population. Schaffendes Volk represented the regime's effort to reconcile individual and collective desires embedded in an emergent consumer culture, with its disciplined and militarized image of a united Reich. The messaging in the exhibition's ideologically explicit displays dissolved the boundaries between public and private spheres. The same modern media techniques were used to satiate the quest for personal identity through consumption, as well as to interpellate the individual into the phantasmagoria of the Aryan racial collective. The split consciousness in the hailing of the consumer can be considered another facet of Nazism's pragmatic reactionary modernism; one which accommodated modern aesthetics and völkisch iconography, allowing the regime to appear both avant-garde and tied to its timeless and pre-industrial roots.

FOUR

WHOSE MODERNITY?

Utopia and Commerce at the 1939 New York World's Fair

"Expositions are the timekeepers of progress."
—*President William McKinley*

AMERICANS enthusiastically embraced the Second Industrial Revolution. For most of them, sometimes despite severe class divisions, the "marriage" of science and technology and the rationalization of industry in mass production marked the advance of the nation. Along with "progress," "modern" was a national keyword. An ethnically and culturally diverse nation, lacking a unifying monarchy or even a long common history, the United States of America looked to a glorious common future. Despite a nod to the founding fathers and the US Constitution, the 1939 New York World's Fair, unlike the other expositions discussed in this volume, virtually ignored the past in its quest to define the future. But recent events had raised questions about the path to that future.

The 1939 New York World's Fair was the scene of a debate over the character of progress. America's deep-rooted love affair with progress—defined largely through science and technology—had survived the ravages of the Great Depression.[1] Despite fears of technological unemployment,

57

most Americans looked for the way forward, the modern way, in organized knowledge—in invention, science-related technology, and agriculture. But progress for what? For whom? And by whom? The 1939 fair, like
many large projects mounted in a democratic society, was the end result
of coalition politics—the product of vigorous debates and only partially
palatable compromises. It was a fair organized in the midst of an ongoing
economic depression and constructed in the face of alarming preparations
for a new world war. This chapter will examine the origins of the fair and
the character of the debates over its message. The fair reflected disagree-
ments over which path to the future America was to take, not only between President Franklin Roosevelt's "New Deal" coalition and business
interests suspicious of it, but also divisions within that coalition itself.

The New Deal was the Roosevelt administration's rapid response to
the devastating economic depression that had crippled America and the
world since late 1929. In the first one hundred days after assuming office in
March 1933, Roosevelt embarked on nothing less than a total reorganization of the economy. The key concept was "national planning." Modernity
was no longer to be defined as freebooting capitalism but, rather, would
include some societal limitations through rational planning. Three key
pieces of legislation were the Agricultural Adjustment Act, the National
Industrial Recovery Act (NIRA), and the Tennessee Valley Authority
(TVA). These acts placed a national plan for production and distribution
for both agriculture and industry at the heart of recovery. The basis of
NIRA was a plan to promote cooperation between trade groups under
federal government supervision, a kind of corporate-government partnership. TVA was a grand plan to modernize a vast, underutilized region.

The *plan* was key. In his campaign for the presidency, Roosevelt proclaimed his commitment to planning: "I plead not for class control but
for a true concert of interests. . . . In this sense I favor economic planning,
not for this period alone, but for our needs for a long time to come."[2] And
later in the campaign: "We need unity of planning, coherence in our administration, and emphasis on cures rather than upon drugs."[3] All within
the New Deal coalition (and even some without) accepted and built upon
the idea of planning. Divisions arose, however, on its meaning. In his book
The Age of Reform the American historian Richard Hofstadter distinguishes
between the "hard" and "soft" sides of the New Deal. The "soft" New

Fig. 4.1 "For a New Deal" campaign button.

Dealers were motivated by idealistic fervor; they wanted to serve "the little people." Some would call them utopians; they were perhaps more appropriately termed "eutopians," or seekers after the "good place." Many were academics, social thinkers, or urban planners. The "hard" New Dealers were the tough-minded, practical (Hofstadter called them pragmatic), get-the-job-done types. Many were lawyers, economists, or businessmen.[4] The social reform phase (the so-called First New Deal) ended around 1935, though the core issues and debate about them remained. The Supreme Court of the United States declared NIRA unconstitutional; TVA under new directorship increasingly turned from being an exciting experiment in regional development to a corporation (as the historian Arthur Schlesinger Jr. termed it) for the production of electric power and fertilizer.[5] As the TVA director David Lilienthal put it: "I do not have much faith in uplift."[6]

But what was the second New Deal to be like? Schlesinger succinctly captures the change: "Where the First New Deal contemplated government, business and labor marching hand in hand toward a brave new society, the Second New Deal proposed to revitalize the tired old society by establishing a framework within which enterprise could be set free."[7] It was over this terrain that the struggles over the message of the 1939 fair would be fought. Already at the 1893 Chicago World's Fair, the historian Henry Adams, the scion of a distinguished political family, believed he discerned America's choice of the future. Modernization was to follow the path of a new "capitalistic, centralizing and mechanical order" able to "create monopolies capable of controlling the energies that America adored."[8] Resistance to this path to modernity provided a tension in American political life that persisted throughout the twentieth century and was exacerbated by the experience of World War I and the Great Depression of the 1930s. The world's fair held on a reclaimed landfill in New York City in 1939 and 1940 provided a playing field for the expression of this strain.

Popular lore has it that the original idea for the fair originated in 1935 with an engineer from the borough of Queens, Joseph Shadgen, who came up with an idea for a fair commemorating the 150th anniversary of Washington's inauguration as president in 1789. He took this idea to Edward Roosevelt, a distant relative of the president. Their concept came to the attention of a group of businessmen who had already been meeting to talk about ways to advance the economic development of the city of New York. The leaders of this group were George McAneny, Percy Straus, and Grover Whalen. McAneny, a banker, was a former Manhattan borough president and the current president of the Regional Plan Association (RPA), a nonprofit organization to stimulate the economic development of the New York metropolitan area. Straus was the president of R. H. Macy and Company, and also a leader of the RPA. Whalen was the president of Schenley Distillers Corporation and as a former city police commissioner was well-connected with city officials. McAneny particularly was immediately taken with the idea of a fair, and assumed leadership.[9] The recent Chicago World's Fair of 1933 provided an instant model.

Chicago's Century of Progress Fair had been a huge financial success. According to Grover Whalen it brought $770 million in new money to the city, retired the bonds by 1934, and even returned a profit to the Fair Corporation. A delegation of the New York business group traveled to

Chicago and returned with glowing estimates—a New York fair would bring over a billion or even a billion and a half new dollars to the City. The group decided at once to incorporate a World's Fair committee with George McAneny at its head.[10] McAneny moved quickly to line up influential local government supporters: Robert Moses, the powerful New York City parks commissioner who wanted to develop Flushing Meadows, an underused part of the borough of Queens; Mayor of New York Fiorello LaGuardia, a rare New Deal Republican; and Governor of New York State Herbert Lehman. Even President Roosevelt expressed interest, saying, "I have been very much interested in hearing of the possibility of an exposition to be held in New York in 1939."[11] Roosevelt had been a commissioner of the 1915 San Francisco Panama-Pacific Exposition and had a continuing concern for the potential of fairs to mobilize public opinion.[12] The First Lady, Eleanor Roosevelt, who was keenly aware of community and homestead developments, also took an active part.[13] The New Deal political leaders were highly aware, as were business leaders, of the importance of public opinion. Fairs such as those of London, Philadelphia, and Paris as well as Chicago's Columbian Exposition, among numerous others, were highly significant in shaping what in the nineteenth century was termed "public sentiment" and in the twentieth, "public opinion." Along with print media, lecture circuits, and evening classes, world's fairs helped set the terms of public discourse on important issues of the day and framed attitudes toward technical change and political choices.

Meanwhile, at the end of 1935 a group who styled themselves as "progressives in the arts" became alarmed lest the coming World's Fair be captured entirely by commercial interests and a great opportunity for public education be lost. The group—mostly architects, designers, artists, and city planners—organized a dinner on December 11, 1935, and invited Lewis Mumford, an architectural critic, urbanist, and public intellectual, to address them on the topic of the proposed 1939 World's Fair.

Mumford's address had a dramatic effect. He argued that most fairs, including the monumentally successful Chicago fairs of 1893 and 1933, oriented themselves toward recounting the achievements of the past; the coming fair should look to the building of a new and better future. It would be a blunder, he said, "to permit it to design itself according to the bids of the highest bidder." He went on to say that "although the industrial story is an important story to tell in the Fair . . . it is by no means the whole story." To tell the "dead story of how wonderful the machine is . . . is not a coherent story and not one which has any educational . . . cultural [or] dramatic value today." What was needed was something very different. "The story we have to tell . . . is the story of this planned environment, this planned industry, this planned civilization. If we inject that notion as a basic notion of the Fair . . . we may lay the foundation for a pattern of life which will have an enormous effect in times to come."[14] The group formed a new committee, the "Fair of the Future 1939 Committee," whose manifesto echoed the December 11 proceedings: "The Chicago Fair looked back over a Century of Progress; the New York Fair should look forward to a Century in the Making. By producing a Fair of the Future, New York will help create the America of the Future. . . . Because Fairs of the past have been retrospective, because their theme has been manufacturers and merchandise and not the social consciousness of . . . new processes and products, they have lacked architectural unity and significance." This manifesto was signed by Mumford, Alfred Barr of the Museum of Modern Art, the designers Henry Dreyfuss, Raymond Loewy, and Walter Dorwin Teague, and the architect Edward Durrell Stone, among others.[15]

The "Fair of the Future" idea caught on across a broad spectrum of political and social viewpoints. There was opposition, of course. Frank Jewett of Bell Laboratories, a moving force behind the Chicago Century of Progress Fair, called the Fair of the Future idea "unconvincing" and "mainly an assemblage of some high hopes and aspirations, coupled with some glittering generalities." Even Fair President George McAneny opined that "there are apparently a good many things in the brief that are debatable." But the Fair Corporation itself lacked a clear and dramatic vision, one that would supersede Chicago 1933. Despite his reservations, McAneny's own interpretation of it, expressed in an address to the Merchants Association of New York in March 1936, made the best of it: "We want . . . to create a spectacle that will be so stunning and impressive that people will come from all parts of the world. . . . It is our hope and expectation that this fair shall celebrate not only commercial progress, not only the progress of inventive genius and its results . . . but that it should celebrate the cultural progress of America, its progress in social and educational directions, in government and administration."[16] Within the Fair of the Future Committee itself designers such as Walter Dorwin Teague, Norman Bel Geddes, and Raymond Loewy could interpret "planning" and "social" as within the broad embrace of planning, but with a major emphasis on the needs and wants of consumers. Mumford and his allies such as Robert D. Kohn and Michael Hare could continue to see the fair's public education role as an opportunity to have an impact on future patterns of life in America. Corporate sponsors and those New York businessmen whose main interest was development could see "planning" and "progress" as partnerships in free enterprise. For example, Edward Bernays, the "father of public relations" in America, advised the Merchants Association that "Such a fair must show graphically the interrelationships of the various groups that make up our life—the relationship of private industry and private enterprise of men and management." Private enterprise, he continued, is the reservoir that is the source of our schools, agriculture, health care, and more. Bernard Lichtenberg, the president of the Institute of Public Relations and an advisor to corporate exhibitors, clearly understood the coming sea change: "Other fairs have been chiefly concerned with selling products; this one will be chiefly concerned with selling ideas."[17]

The New York Fair of 1939 brought into sharp relief changes in industrial displays at world's fairs that had been taking shape since the end of the nineteenth century. Most of the fairs of the previous century emphasized the presentation of products. By 1915, the year of the San Francisco Panama-Pacific Exposition, the leadership of major corporations began to highlight the processes of production, reinterpreting the technical education mission of fairs as revealing the magic and majesty of industrial production in a systematic way. At Chicago in 1893, nine of 137 buildings were constructed by individual companies. Forty years later, at the Century of Progress Fair, twenty corporations, including some of America's largest companies, had their own buildings.[18] At the 1933 Chicago Fair, some corporations in their zeal to promise that science via corporate research would solve the problem of the Great Depression began to shift toward highlighting research rather than the details of production. At the 1939 fair, using showmanship and taking their cues from the new professions of advertising and public relations, corporations such as General Motors, General Electric, and Ford began to suggest that, as General Electric's advertising much later phrased it, "progress is our most important product."[19]

The official theme adopted for the 1939 fair was "Building the World of Tomorrow with the Tools of Today." Mumford ally Robert D. Kohn was named chairman of the theme committee; he sought to advance the Fair of the Future program through public outreach. A successful New York architect, Kohn was a member of the Regional Planning Association of America, along with Mumford, Clarence Stein, and many others influenced by Ebenezer Howard and Patrick Geddes. He had been a founding member of the Technical Alliance formed around the New School in New York in 1920 and composed of followers of the economist Thorstein Veblen. Members of the organizing committee of the Alliance included Kohn, Benton Mackaye, Stuart Chase, Louis Comstock, the architects Frederick Ackerman and Charles H. Whitaker, engineer-scientist Charles Steinmetz, engineer Bassett Jones, physicist Richard Tolman, economist Leland Olds, and several others.[20] Ackerman, Jones, Whitaker, and Comstock joined Kohn in developing the 1939 fair. Comstock, an electrical engineer, designed the fair's lighting; Jones was head lighting consultant.

In an address on the *New York Times*'s radio station WQXR late in 1936 Kohn outlined his vision: "The seeds of great changes lie all around us today. They are found in the streamlined train, the airplane, the modern house. They are also found in less tangible things as new forms of organization in government and business and new techniques like proportional representation and cooperation." Focusing on his theme of "interdependence," Kohn argued that inventions tie us together. So, at the fair, "instead of separate buildings for Art, Science, Industry and so on . . . the visitor to this Fair will find every exhibit arranged . . . according to the different features of his ordinary life. . . . The focal exhibits in each field will form a natural gateway to the . . . products and processes which fit logically into that field."[21] In an article published just before the opening of the fair, Kohn reiterated:

> We are always showing [the people] things in isolated categories. Why not exhibit the forces that connect them—the living ideas that alone make things useful or harmful? Accordingly, the Fair designers could not divide it into such categories as science, art, agriculture, manufactures— the classic divisions convenient for technicians but not illuminating to laymen. We chose to make our major divisions more or less functional,

the things with which the average man comes in contact in his everyday life. . . . What is more, instead of isolating science and art, the planners would attempt to show them permeating all of these other things, as illustrations of their interpenetration into the functions of modern life.[22]

By mid-1937 the theme diagram evolved to look like the following plan. Dominating the central space would be the symbolic heart of the fair, the 610-foot-high three-sided pylon called the Trylon, next to a gigantic 185-foot-diameter hollow sphere called the Perisphere. Inside the Perisphere would be mounted the theme exhibition *Democracity*. To the right of the Trylon and the Perisphere would be arrayed exhibit buildings devoted to production, distribution, and nourishment. To the left would be found means of communication, business administration, clothing, and "community interests." Directly in front lay the Government Zone housing the large Federal Government Pavilion and those of foreign governments. To the south of the Trylon and the Perisphere would be the large transportation zone.[23]

In 1936 George McAneny stepped down as president of the Fair Corporation and was replaced by Grover Whalen, former New York City police commissioner and head of Schenley Distillers. Whalen was a promoter and a New York City enthusiast and its former police commissioner, whose interest was less in social education than in consumer education. It was Whalen who led the fair team to Washington, DC, to testify before the House of Representatives in favor of a bill authorizing federal participation in the fair and funds amounting to $5 million. On March 23, 1937, Whalen provided the now-official line of the theme committee:

> We were not satisfied to create merely another fair. We realized we were
> in an era of social and economic readjustment: we could not—we dared
> not—present a smug picture of easy contentment; we felt it necessary
> to bring to the people of the world a story of the hopes which lie in our
> children's tomorrow. . . .
>
> By setting forth what has been beside what is, the fair of 1939 will predict, may even dictate, the shape of things to come . . . [The visitor may]
> gain a vision of what he might attain for himself and for his community
> by intelligent, cooperative planning toward the better life of the future.[24]

President Roosevelt drafted an endorsing letter later that year, stating, "The nations of the world are accepting the exposition as a medium for the advancement of better international relations and for the establishment of a fuller understanding and appreciation of the complex life of our day."[25]

Whalen brought with him to Washington an impressive array of elected officials, academics, businessmen, and labor leaders. Matthew Woll, vice-president of the American Federation of Labor and a member of the executive committee of the Fair Corporation's board of directors, testified on behalf of organized labor. Labor itself had no pavilion, but, Woll argued, organized labor had participated in the planning and would benefit from the results. Woll estimated that two hundred thousand workers were finding employment in the planning, preparation, operation, and maintenance of the fair; another eight hundred thousand would "be indirectly engaged." "But we come," he continued, "not as a selfish group with a selfish motive." Rather, "we feel in 1939 there will be a decisive change and the cry in Europe and everywhere will be 'Let us come to America' . . . Thus everyone will benefit from this, the greatest advertisements of all." Woll indicated that labor was brought into participation in the planning from the beginning, and was well represented in the councils of the fair. "We therefore glory in this whole enterprise."[26] The resolution passed easily.

The year 1937, however, witnessed renewed economic stress in the United States. A severe recession threatened a major deepening of the Depression. Within the world's fair enterprise the hand of the corporate sponsors was strengthened. As budgets tightened, corporate sponsors of exhibit space greatly increased their role. The soft New Dealers, like Mumford, Hare, and Kohn, who advocated turning the fair into a vehicle for popular education and social reform, gave ground to hard New Dealers, to experienced corporate hands, and to industrial designers like Teague, Bel Geddes, and Loewy, who aimed at pioneering new and dramatic forms of presenting the corporate world's technological and commercial progress.[27]

In March 1939, just before the opening, the theme committee issued a press release to appear in the *New York Times* that clearly illuminates the ground ceded by the social reformers:

It might well be presumed that the first and most vigorous response to this forward-looking program would come from those engaged in purely intellectual pursuits, the educator, the civic planner, etc. But it was the so-called hard-boiled business man and industrial leader who reacted most enthusiastically. . . . The practical idealist's version of the important part machinery will play in the scheme of things in the future . . . is presented in a vivid manner by the model City of Tomorrow Morning being fashioned as a principal feature of theme spectacle of the New York World's Fair 1939. . . . Cities in the past . . . were permitted to grow at the pace of the horse. The City of Tomorrow Morning will be inspired by the automobile and the speed made possible by improved traffic arrangements. Therefore the community will speed outward rather than grow upward in the congested fashion of the city of today.[28]

The opening day of the fair was April 30, 1939. Arrayed in front of the symbols of the theme center, the Trylon and Perisphere , the audience heard the luminaries of the New Deal, President Roosevelt, Governor Lehman, and Mayor LaGuardia, the last of whom spoke of a fifth freedom —social security—to strengthen Roosevelt's famous four.[29] Wind and rain dampened the festivities and "by five o'clock the flow out of the fairgrounds had become a rout. . . . Even New Yorkers, it seemed, have not learned the secret of being happy in wind and drizzle—not even in the World of Tomorrow."[30]

The actual layout of the fair adhered roughly to the functional schematic of 1937 described above, but instead of an education in the interrelationships of the elements that compose our daily lives, visitors to the fair were more likely to be introduced to the latest technologies and the most attractive products. Visitors would enter through the Corona Gate into the transportation zone illustrating mankind's conquering of time and space, which included aviation, marine transportation, and railroads but yielded pride of place to the automobile manufacturers General Motors, Ford, and Chrysler. Crossing via the Bridge of Wheels or the Bridge of Wings the visitor would encounter the New York City pavilion situated next to the Trylon and the Perisphere, which contained the theme exhibition, *The City of Tomorrow,* or *Democracity.* To the left, as planned in the 1937 schema, was the communication and business zone, to the right the production and distribution zone, which included the pavilions of General Electric, Westinghouse, US Steel, Dupont, Consolidated Edison (New

Fig. 4.2 Trylon and Perisphere. Gottscho-Schleisner Collection, Library of Congress.

York's electric and gas company), and industry-wide exhibition buildings such as petroleum, men's apparel, metals, and pharmacy.

Reaching beyond the theme center lay Constitution Mall, which led into the Lagoon of Nations and the vista beyond it to the Court of Peace and the grand pavilion of the United States Federal Government. To the left of the mall was the community interests zone; to the right, the food zone. Surrounding the Lagoon of Nations, in the government zone, were arrayed the pavilions of the many nations recruited for the fair and smaller exhibitions in the Hall of Nations. Cuba was placed in the amusement zone, Sweden and Turkey were situated in the food zone, and Jewish Palestine in the community interests zone. There were no public explanations for these anomalies.

By the time of the opening of the Fair, international tensions had taken their toll. Even before the Czechoslovakia building was constructed, the nation ceased to exist as an independent entity; before the fair ended in 1940, Poland, too, vanished. Several major powers did not participate. China was under siege by Japan; Spain was ravaged by civil war; and Germany was hesitant about demonstrations and public opposition to its policies by Mayor LaGuardia and New York City's anti-Nazi citizens. The official excuse was the problem of foreign exchange.

Great Britain made the most adroit use of its participation for propaganda purposes. Its aim was clear: the British Foreign Office wished to promote Anglo-American unity in the face of a coming conflagration. Despite President Roosevelt's pro-British sympathies, sentiment in the United States had a strong isolationist component. A campaign to underline commonalities between the two countries was begun in earnest. The British government readily endorsed Roosevelt's invitation to King George VI and Queen Mary to visit.[31] The pavilion itself was designed to appeal to the American public and promote several key messages. The foremost concept was the two nations' common democratic heritage. The "Hall of Democracy" featured an original copy of the Magna Carta, which newsreels touted as the "basis of modern democracy" and "the common bond linking England and the United States." After war broke out, the document was given to the Library of Congress for the duration of the war for "safekeeping."[32]

France's contribution was a well-sited modernist building that appeared to be devoted mainly to making France interesting and attractive to visitors. The entry level was designed to encourage tourism; it contained scenery and interiors of homes from four French provinces. The mezzanine displayed French fashion (hats, gowns, and perfumes), and examples of the fine arts such as painting, sculpture, and tapestries. The upper floor was devoted to other aspects of French culture (literature, furniture design, French history through art) and one of the most visited and commented-upon attractions of the fair, the French restaurant. The restaurant lived on after the fair in the form of the famous New York City French restaurant Le Pavillon.

Japan's exhibit at the fair was an exercise in suppressing context. Japan's relationship with the United States was deteriorating as it prosecuted

Fig. 4.3 Italian pavilion. Gottscho-Schleisner Collection, Library of Congress.

a very bloody war against China, a nation that commanded much sympathy in America. The exhibit was modeled after an ancient Shintō shrine, surrounded by beautiful gardens. Attractive young women in traditional costume gave talks about Japanese culture. A simple stone lantern held a flame, the "flame of peace." On April 2, 1939, Japan's premier, Baron Hiranuma, made a goodwill broadcast in honor of the fair on all American radio networks. He said that relations between Japan and the United States had been growing closer since 1854. "It is my firm belief," he said, "that our neighborly relations so happily begun and so well fostered will continue to grow in strength through a mutual understanding."[33] The Diplomatic Room provided a record of these friendly relations and featured a replica of the American Liberty Bell made out of thousands of cultured pearls and silver, and valued at over $1 million.[34]

Grover Whalen exerted special effort to recruit Italy's participation, visiting Mussolini in Rome. Il Duce responded with gusto; the Italian pavilion provided a portrait of his vision of the New Italy as a revived, high-technology version of the Roman Empire. Italy's Commission General for the Fair, Admiral Giuseppe Cantu, expressed his country's aims quite clearly in an article for the *New York Times*. "Italy's participation in the New York World's Fair," he wrote, "aims particularly to acquaint the American public with the New Italy. Not the legendary one of the history books . . . but the Italy which has emerged during the last fifteen or sixteen years with her highly developed industries, her great hydroelectric plants, her renascent handicrafts, her modernized agriculture and many other industrial activities which today place her in the forefront of European nations." Cantu went on to describe Italy's own world's fair, "The Olympiad of Civilizations" planned for 1942, represented in New York by a small-scale model in the Tourist Hall of the Italian pavilion.[35] The pavilion itself was a neoclassical palace. The goddess Roma was placed atop a two-hundred-foot tower, and from the base of her pedestal a waterfall cascaded into a pool featuring a monument to Guglielmo Marconi, one of the early inventors of radio. The first floor was dedicated to demonstrating Italy's technological self-sufficiency ("Autarchia") through the display of machines to produce synthetic fibers, aeronautics, naval prowess, and electro-technology. An entire exhibit was devoted to the achievements of Marconi. An imposing statue of Mussolini adorned the Italian part of the Hall of Nations, outside the pavilion.[36]

One of the largest and most expensive pavilions at the fair was that of the Soviet Union. Boris Iofan, the architect of the Soviet Union's imposing 1937 Paris Exposition structure, was named chief architect. As in the 1937 Paris Expo, daunting size seems to have been one of the important motifs of the Soviet exhibition. At the earlier fair, a gigantic verti-

Fig. 4.4 Soviet pavilion. Gottscho-Schleisner Collection, Library of Congress.

cal façade supported Vera Mukhina's huge statue of a worker and peasant woman holding up a hammer and sickle. In New York, visitors met a 79-foot stainless-steel statue of a worker holding a red star, both situated atop a 180-foot pylon. The entrance lobby sported a 53-by-30-foot mural depicting Soviet people happy in their lives under socialism. Reiterating the premise of progress since the revolution—industrially, socially, and politically—eight exhibits illustrated all aspects of Soviet life. Underlining the theme of the World of Tomorrow, there was a hall devoted to socialist city planning, and a detailed exhibit of a ten-year plan to refashion Moscow. The reception by visitors and the press to the Soviet pavilion was generally positive. The *New Masses* (affiliated with the Communist Party USA) termed the pavilion "brilliant, true to the material used, the nature of the Fair and the intention of the exhibits." Still in Popular Front mode, the *New Masses* and the *Daily Worker* generally approved of the fair: "The industrial exhibits are nice, but too much has been left unsaid. The spirit of international cooperation, however, we like, and we hope that out of this advance edition of the World of Tomorrow will come more than the humbug Grover Whalen is spreading."[37] Unsurprisingly, the left-wing press was enthusiastic about the Soviet exhibition, but *Time* magazine, too, ranked the Soviet effort as the best foreign exhibit at the fair.[38]

The foreign pavilions added color and contrast to the fair, but by far the best-attended and most faithful to the expressed themes of the fair were the United States' homegrown government and corporate exhibits. It was here that the inherent tensions in the interpretation of modernity made themselves manifest. As concrete examples of how the dueling visions worked themselves out in the fair, we can turn to three visions of the American city of the future: Henry Dreyfuss's *Democracity*, General Motors' *Futurama* (the fair's most popular exhibit), and the film titled *The City*, produced for the fair with a script written by Lewis Mumford.

Democracity, or the *City of Tomorrow*, was housed in the fair's theme center within the eighteen-story Perisphere. It was designed by the industrial designer Henry Dreyfuss under the supervision of the chairman of the theme committee, Robert D. Kohn. Before the fair opened, Kohn described it in a manner resembling the Garden City notions of Ebenezer Howard and with a nod toward the Regional Planning Association of America of Clarence Stein, Henry Wright, Mumford and Kohn himself: "Democracy in the World of Tomorrow. . . . Not a dream city but a symbol of a life to be lived by men tomorrow. . . . [T]own and country joined for work and play in sunlight and air. A brave new world built by legions of hands and hearts working as one."[39] As the official pamphlet urged, "Democracity dramatizes the meaning of the Fair. . . . The meaning is simple: Consciously or not, we are building the World of Tomorrow; creating the symbols of living; not each for himself, but all together. Hence the double

Fig. 4.5 Entering *Democracity* in the Perisphere. Gottscho-Schleisner Collection, Library of Congress.

theme of the fair . . . building the World of Tomorrow . . . and the interdependence of man."[40]

Visitors entered the Perisphere via escalator and took their places in one of two rotating balconies. For about six minutes, to the accompaniment of music composed by William Grant Still and the narration of newscaster H. V. Kaltenborn, the audience revolved around what was represented as a region of 2039. "Centerton," with its unique skyscraper, was the commercial center of the region, surrounded by satellite towns such as "Millville" (an industrial town), farm communities, and "Pleasantvilles," or residential suburbs. The entire region, dotted with green spaces, was populated by a million and a half residents, with Centerton accounting for 250,000 of them.[41] A TVA-like dam provided hydroelectric power and sleek motorways connected Centerton to the rest of the region.[42]

Reflecting Kohn's earlier draft, Kaltenborn intoned: "The City of Man in the World of Tomorrow. . . . Not a dream city but a symbol of life as lived by the Man of Tomorrow. . . . [T]own and country joined for work and play in sunlight and good air. A brave new world built by united hands and hearts. Here brains and brawn, faith and courage are linked in high endeavor."[43] This vision of *Democracity* emphasized the regional approach to urban design but even in 1939 seemed a sanitized version of the present. In the *New Yorker* magazine, Lewis Mumford accused it of presenting "a stale conception of the capital city," which is "masquerading as the future."[44] Art historian Francis V. O'Connor describes it as "a planner's version of regional suburban sprawl."[45] The RPA, however, saw it as "suggestive of the power for good that is inherent in city planning."[46]

Less than half a mile away, the General Motors pavilion, officially titled *Highways and Horizons*, included a different view of the future—one that was touted as "a magic Aladdin-like flight through time and space" or *Futurama*. Created by Norman Bel Geddes, a former set designer turned industrial designer, the audience was drawn from a pristine world of hills, valleys, and farms into a world fashioned by mankind's "industry and genius" via new, great motorways toward a great towering city, "The City of 1960." Mumford, in the *New Yorker*, declared, "Here is a town that brazenly repeats the bad planning of Gary, Indiana, or Lackawanna, Pennsylvania, by placing workers right under the dirt and fumes of the steel plant, within walking distance."[47] But *Futurama* was the fair's most popular exhibit by far; people queued for hours to enter it. Waldemar Kaempffert, the science writer, claimed that "the best exhibit at the Fair is the train and sound system [of Futurama]. And both, paradoxically, are not exhibits at all."[48]

Bel Geddes, the designer, convinced General Motors to provide a way of making the visitors share an experience that would not, as previous exhibits had done, attract them by demonstrating the intricacy of industrial production but by having them participate in the company's vision of its social role and its place in the future. The GM brochure accompanying *Highways and Horizons* claims that "when you visit the Research exhibit you see actual methods by which Science is steadily advancing the cause of Progress. . . . From all this it is our hope that we make clear the enduring purpose of General Motors, which is not merely to make cars and refrigerators and Diesel engines and the like for today, but constantly to promote the welfare and progress of the nation."[49] As historian Roland Marchand put it, GM would "entice visitors not to 'tour our factory' but instead to 'share our world.'"[50]

Visitors entered a dimly lit entrance hall and then the main hall itself. The audience was seated by twos in moving chairs divided into "cars" with their own speakers so that what they saw corresponded with the narration. The moving chairs had "wings" strictly controlling the line of sight. They were carried over a fourteen-lane motorway at varying high speeds, traversing farms, valleys, mountain vistas, and suburban landscapes (all rendered with spectacular detail), toward the future city. Night would fall, and lighting effects would create a different audience experience. Finally, they were brought into the new city in which two-level streets separated pedestrians and traffic. The visitors looked upon the street from above, but suddenly they would be swung about and enter the full-size street at ground level. The visitor would get out of the chair and walk about.[51] All would get a pin with the message "I Have Seen the Future." Over twenty-seven million people braved waits of up to two hours to see this version of the future.[52]

The typical audience reaction was perhaps best captured by the extensive *Life* magazine article "*Life* goes to Futurama" that appeared on June 5, 1939. According to *Life*, in 1960, "The highways skirt the great cities . . . but the happiest people live in one-factory farm-villages producing one small industrial item and their own farm produce." On the motorways, "the cars built like raindrops are powered by rear engines. . . . Inside they are air-conditioned. They cost as low as $200." The "city of 1960 is inspired by 1500-ft skyscrapers, widely spaced with parks on the roofs of the intervening low apartments."

But the key is the scientific-technical world that lay beneath the surface.

> Behind this visible America of 1960, hidden in the laboratories, are the inventors and engineers. By the spring of 1939 they had cracked nearly every frontier of progress. . . . Liquid air is by 1960 a potent, mobile source of power. . . . Cures for cancer and infantile paralysis have extended man's life span and his wife's skin is still perfect at the age of 75. . . . All of this—much of it to be seen in the model in the General Motors Futurama at the New York World's Fair—is a vision already conceived by 1939's engineers.[53]

Fig. 4.6 Street level intersection in *Futurama*.

For the *New Yorker*, the essayist E. B. White remained skeptical: "A ride on the Futurama of General Motors induces approximately the same emotional response as a trip through the Cathedral of St. John the Divine. . . . [You] hear . . . the soft electric assurance of a better life—the life which rests on wheels alone—[You go] a hundred miles an hour around impossible turns ever onward toward the certified cities of the flawless future. . . . [But] where will the little bird build its nest?"[54] Walter Lippmann, another respected public intellectual of the time, came to another, somewhat paradoxical conclusion, one that resonated with the New Dealers of the second New Deal. He reported that *Futurama*

> is as proud an exhibit as one could find of what men can achieve by private initiative, voluntary organization, individual leadership and the

personal genius of scientists and inventors. . . . [But] one realizes that this paradise of the motorist will have to be constructed not by private enterprise but by a public works administration. General Motors has spent a small fortune to convince the American public that if it wished to enjoy the full benefit of private enterprise in motor manufacturing it will have to rebuild its cities and highways by public enterprise. Soon one comes away feeling . . . that both are necessary and that their collaboration is indispensable.[55]

A third vision of the city of the future presented at the fair was intended as something of an antidote to the regionalist but much compromised *Democracity* and the corporate vision of *Futurama*. It was a specially produced forty-four-minute film titled *The City* that was shown at the fair's Science and Education exhibit. Its origins lay in an idea hatched by Robert Kohn and planner Catherine Bauer at least as early as 1937. "Miss Bauer and I," Kohn wrote, "have been working for a long time . . . to make a movie illustrating the way an American city has grown in the past (planless) and on the same site how it might grow under a carefully conceived plan." A committee for making the film was organized with Frederick Ackerman, the technical director of the New York City Housing Authority; Tracy Augur, the president of the American Institute of Planners; and Robert Kohn and Clarence Stein of the now defunct Regional Planning Association of America. The committee morphed into a company created for the purpose called Civic Films, Inc., with Stein as president and Ackerman, Augur, and Kohn as directors. The original idea was to recruit as composer Virgil Thomson (who wrote the score for *The Plow that Broke the Plains*) and as narrator "someone of the type of Archibald MacLeish." In the end, Ralph Steiner and Willard Van Dyke were the directors, with a script written by Lewis Mumford based on an outline by Pare Lorentz (*The River, The Plow that Broke the Plains*). A young composer named Aaron Copland wrote the music and the narration was presented by the actor Morris Carnovsky. A grant of $50,000 from the Carnegie Corporation provided the main funding.[56]

The film begins with a lost Eden—a New England village. Nature and technology exist in harmony. Water wheels, a village smithy, and barrels being made by hand provide a glimpse of technology in balance. Suddenly, the Industrial Revolution intervenes: smokestacks, blast furnaces, children playing in mud, tenements. Cacophonic music underlines the disharmony of technology and nature. Is this the future? Suddenly . . . calm. We enter a new world via a sleek modern airplane. We enter the world of *neotechnics*, or modern technology. Clean water, grass, and children on bicycles underline the contrast. The lost Eden is restored by good planning and good technology.[57]

The film was shown eight times a week at the fair and generally received good notices if not huge crowds. In an unusual self-review in the *New Yorker*, Mumford describes the film (probably accurately) as "a belated attempt at salvage," but, he continues, "[it] leaves off . . . at the point where a new demonstration should begin."[58] Though the crowds were small, critical reaction was enthusiastic. Archer Winsten, critic for the *New York Post*, for example, wrote that "If there were nothing else worth seeing at the fair, this picture would justify the trip and all the exhaustion." For this critic, "a new horizon has come into clear sight."[59]

Winsten's reference to a new horizon relates to the General Motors exhibition's official title, *Highways and Horizons*, and to its accompanying film *To New Horizons*. This promotional film for the company provides a visual tour of *Futurama* and in a curious way parallels *The City*. Its accompanying narrative stresses "new horizons in the spirit of individual enterprise." The story tells of progress through better transportation linked to "better living" and more generally to "highways of human activity."[60] While the past and present are shown in black and white, the future is displayed in glorious Technicolor. Unlike *The City* but like *Futurama* itself, *To New Horizons* is the story of monotonic, untarnished, and inevitable progress.

A third film at the fair that attracted considerable attention was *The Middleton Family at the New York World's Fair*, a promotional film entirely in Technicolor by the Westinghouse Corporation. Both *The City* and *To New Horizons* were intended to be inspirational and to appeal to the high-minded. The Westinghouse film was aimed at a "family" audience, to tell a folksy story while creating a demand for electrical products. *The Middleton Family* features an Indiana family's visit to the fair, though all the action takes place in or near the Westinghouse pavilion. The plot involves the young adult daughter, Babs, who has a foreign-born suitor named Nicholas Makaroff and a hometown admirer, a Westinghouse engineer, named Jim Treadway. Makaroff, a college teacher of abstract art, is clearly anti-capitalist, somewhat exotic, and without a sense of humor. Jim is sensible, straightforward, and convinced of progress through corporate enterprise. Slowly, Makaroff is revealed as a fraud and Babs, in the end, chooses Jim.

In Sinclair Lewis's famous novel *Babbitt*, the eponymous hero's "God is modern appliances."[61] *The Middletons* almost seems an extended exercise in babbittry. Throughout the film, the audience is given a grand tour of the Westinghouse exhibit. One highlight is the "Battle of the Centuries," a competition between a Westinghouse automatic dishwasher watched over by Mrs. Modern, and a hand-washing Mrs. Drudge. In the film, as at the fair, Mrs. Modern emerges, rested and composed, as the victor. Mrs. Drudge, harried and disheveled, always loses. This little drama was staged forty times a day with eight interchangeable Mrs. Moderns and Mrs. Drudges.[62]

In the end, the 1939 New York World's Fair sent multiple messages. Under the broad umbrella of "working together, we can build a better world of tomorrow," the New Deal strategy of harnessing public and private enterprise found its outlets, as did the corporate emphasis on individualism and free enterprise. But all embraced the theme of organized knowledge for a better future. Science, technology, research, and planning for a better future were tropes that all participants agreed upon. However, just as the fair opened its gates, developments in Europe and Asia seemed to shred its presuppositions and overshadowed its messages. By June 1940 twelve of the foreign countries at the fair were drawn into the European war, and US-Japanese relations were rapidly deteriorating. The League of Nations pavilion remained a sad reminder of its ineffectiveness. The *New Yorker* sarcastically referred to the fair as "The World of Yesterday."[63] The slogan, "Building the World of Tomorrow," was replaced in 1940 by "The Fair for Peace and Freedom." Over its two seasons the fair had (paid) attendance of over forty-five million. It amassed a deficit of almost $19 million.[64] When the fair closed in October 1940 its remains as scrap metal were recycled into America's defense effort in preparation for what many saw as the inevitable war.[65] After the attack on Pearl Harbor, President Roosevelt announced that "Dr. New Deal has been replaced by Dr. Win the War," and the 1939–1940 New York World's Fair passed to the historians.

Twenty-five years later New York mounted another world's fair, the 1964–1965 New York World's Fair, spearheaded by Robert Moses, the city parks commissioner and chairman of the powerful Triborough Bridge and Tunnel Authority. Moses had wanted the earlier fair to be a springboard for development of the borough of Queens, but fair financial losses, the state of the economy, and the coming of the war ended that dream temporarily. He returned to his plan a quarter of a century later. The 1964 fair was held on the same grounds, Flushing Meadows, and this time Moses projected seventy million visitors, a figure that was unmet. Although not sanctioned by the Bureau of International Expositions (and therefore it had greatly reduced international participation) this fair touted itself as a "universal and international exposition." If the 1939 fair could be criticized for not fully living up to its own ideals (and it was), and many exhibits could be faulted for overzealously fostering consumerism (and they were), the 1964 fair paid little heed to ideals. Opened only six months after the assassination of President Kennedy and the end of his vision of the "New Frontier," the subtheme of "new frontiers" governed this world's fair. The General Motors 1964 exhibit *Futurama II*, for example, can profitably be compared to its 1939 *Futurama*. In 1939 visitors were transported to the city of *their* future, presented to them as something mankind (especially through its corporate tribunes) could build with the tools of today. *Futurama II* presented visitors with a glimpse of how future technologies would conquer new frontiers of deep space, dense jungles, the deep ocean, and desert landscapes, rendering them habitable and bringing them well within a world market economy. They "visited" a lunar colony; the oceans where undersea "aquacopters" and atomic-powered submarines would drill for oil and harvest the sea; a thick jungle through which vast highways would be built by robot factories chewing up the terrain with laser-beam clippers and laying the foundation for four-lane superhighways, chemically cleared of insects and other pests; deserts made to bloom with atomically desalinated sea-water; and finally, the City of Tomorrow with huge superskyscrapers, multilevel roadways, and covered moving walkways bringing customers to vast shopping areas "which are now truly marketplaces of the world."[66]

At the 1939 fair, the theme that resonated with the expectations of both presenters and visitors was modernity through rational planning. By 1964 the path to modernity already discerned in 1893 by Henry Adams—the "capitalistic, centralizing and mechanical order" able to "create monopolies capable of controlling the energies that America adored"—was clearly ascendant.[67]

FIVE

MODERNITY ON DISPLAY

The 1940 Grand International Exposition of Japan

THIS chapter examines preparations for the Grand International Exposition of Japan that was to have been held in Tokyo and Yokohama over the period March–August 1940 to celebrate the 2,600th anniversary of the mythological ascension of the Emperor Jimmu in 660 BCE. The outbreak of war with China in 1937 forced the Japanese government to cancel the event. Japan's participation in international expositions in the late 1930s and the aborted plans for Japan's own event throw light on the tensions between the promotion of modernist architecture and emperor-centered ideology. Japan was eager to achieve a type of self-sufficiency by expanding its empire and embracing science, technology, and industrialization. At the same time, it flagged the importance of tradition. The compromise that resulted saw modernity combined with a type of romantic nationalism (or perhaps more appropriately labeled Pan-Asianist Internationalism), what Jeffrey Herf calls "reactionary modernism."

Why write the history of a nonevent? In Herf's original study, *Reactionary Modernism: Technology, Culture, and Politics in Weimar and the Third Reich,* modernity, as seen in technology, is a major area of focus.[1] This chapter tests and interrogates the applicability of Herf's concept of "reactionary modernism" by examining the plans for the aborted 1940 exposition and Japanese representation at the 1939 New York World's Fair. The Japanese organizers of the 1940 Grand International Exposition sought to emulate other major international expositions.[2] Despite the mythmaking about the unbroken line of Japanese emperors, Japan did not reject modernity. Rather, a discourse was developed that converted technology from being merely a component of Western civilization to one that was more organically part of Japanese culture and its tradition of making things. This extended to examples of major modern architecture completed in the 1930s, albeit with Japanese inflections. The 1940 expo would have showcased modern technology within the context of the 2,600th anniversary celebrations, reinforcing Japan's commitment to technological progress and incorporating modernity within Japanese cultural nationalism. If it had been held, it would have reinforced the imperialist project of Japan. After the war, the struggle to deal with Western-inspired modernity continued to be played out in Japanese efforts to embed ideological elements in Japanese architecture to differentiate it from modernism elsewhere. Some Japanese architects sought to infuse with cultural traditions what was otherwise seen as an international style.

Science, Technology, and Empire

From the mid-1920s to the mid-1940s in Japan, science was promoted in popular magazines as "a commodity packed with the sense of wonder."[3] This was achieved by pictures of nature and scientific activities involving the observation of nature, experiments, and the gathering of specimens. The popular science culture, this world of wonder, was easily co-opted by a nation that was increasingly on a wartime footing. Magazines excited their readers with articles and stories of weapons and war, encouraging them to be ideal imperial subjects who were "scientifically and technologically capable, as well as loyal to the emperor and the nation."[4]

In the 1930s Japan enjoyed an average annual GNP growth of 5 percent, especially in metals, chemicals, and engineering. The economy was increasingly militarized and propaganda emphasized the emperor cult, but the number of science and engineering graduates during the Pacific War (1941–1945) was actually triple what it had been ten years before. Throughout World War II, Japanese scientists and engineers continued to pursue cutting-edge research and to develop technology.

The fascinating question is how they were able to do this despite a romantic ideology of an unbroken line of emperors and the notion of the emperor as a god. A current of reactionary modernism allowed Japanese to combine the irrationalism of the emperor cult with an enthusiasm for science and technology. The Grand International Exposition was aimed at creating a world of wonder where Japanese imperial subjects and other international visitors could not only be enthralled by the achievements of modern Japanese science, technology, and industry but also celebrate tradition and the imperial line. This was in line with wartime discourse in the mid-1930s and 1940s that argued that Japanese technology was not merely a component of Western civilization but one that was more naturally part of Japanese culture and its tradition of making things.

Mobilizing Support

The story begins with a group of local businessmen and government representatives from Tokyo and Yokohama who came together on May 23, 1930, and established a committee with the aim of holding an international exposition. Two and a half years later, on October 10, 1932, a resolution was passed unanimously to hold the exposition in Tokyo in 1940 to coincide with the 2,600th anniversary of the accession to the throne of Emperor Jimmu. It was also decided that foreign nations would be encouraged to participate. On May 31, 1934, the Association of the Japan International Exposition was created to take charge of the event. The Mayor of Tokyo, Ushizuka Toratarō, was elected president. The Mayor of Yokohama, Ōnishi Ichirō, and the industrialist Hoshino Seki were both appointed vice-presidents. At the Sixty-Seventh Session of the Imperial Diet in 1935, a resolution was passed endorsing the committee's decision. The Tokyo Municipal Assembly also passed a similar resolution.[5]

The Grand International Exposition of Japan was announced to the Japanese people via a radio broadcast on February 11, 1935, by Mayor Ushizuka in his capacity as inaugural president of the Association of the

Japan International Exposition. Ushizuka explained to the public that the exposition would be held in both Tokyo and Yokohama from March 1940 for a period of six months. The date of the announcement itself was an auspicious one, falling on the annual national holiday of Kigensetsu, in honor of the first accession to the imperial throne in 660 BCE that the exposition would celebrate.[6]

In May 1936, the Imperial Diet approved an appropriation for supervision of the preparations for the exposition. A few months later, on August 25, 1936, the cabinet agreed that the Association of the Japan International Exposition should promote the event. An Exposition Supervision Section in the Ministry of Commerce and Industry was subsequently created to assist in the project, and the Minister of Commerce and Industry was appointed honorary president of the exposition on October 7, who was then succeeded by Vice Admiral Godō Takuo when he assumed office as minister on February 8, 1937.[7] Vice Admiral Godō recommended the appointment of Fujihara Ginjirō as president of the organizing body.

Fujihara had just published the book *The Spirit of Japanese Industry* (1936), in which he sought to marry Japan's modernity with its traditions. These ideas had been gestating for some time and Fujihara was not the first to promote such ideas. The book was a translation of *Kōgyō Nihon seishin* (1935), which was in turn based on a series of lectures that Fujihara had given in 1934. As he stated in the foreword to the English version, "It was written for my own countrymen, whom I wished to inspire with confidence in the future of Japanese industry."[8] The text reflects Japanese ideas about the relationship between technology and culture among Japan's business elite. Fujihara was chief executive of the Oji Paper Company (which, among other things, produced newsprint) and prior to that held positions in the *zaibatsu* Mitsui and Company. He was made a life member of the House of Peers for his distinguished service to the nation.

Fujihara sought to reconcile the rapid development of Japanese industry with "racial traits" such as mental discipline, "honor above life for artists and craftsmen," samurai spirit, the relationship between "technical achievement and the Japanese spirit," diligence, a simple lifestyle, and "excellent administration." In addition, the growth of the Japanese empire provided, in his words, "splendid" opportunities for industrial expansion.[9] Fujihara's appointment was followed by other very high-profile ap-

Fig. 5.1 Expo Poster showing kimono-clad woman and Mount Fuji in the background. "International Exposition of Japan in 1940 to Be Held at Tokyo and Yokohama to Commemorate 2,600th Year of Founding of Japanese Empire," *Japan Trade Review* 11, no. 2 (March 1938): 38.

pointments when Konoe Fumimaro became prime minister in June 1937. Prince Chichibu was appointed president of the exposition, Konoe became vice-president, and Minister of Commerce and Industry Yoshino Shinji became an honorary chairman, along with Baron Sakatani Yoshio. These key political figures were all keen to be involved and to lend their credibility to an event that was estimated to attract forty-five million people.[10]

1940 Grand International Exposition and the Olympics

The original plan was to schedule the Olympics in Tokyo in 1940, directly after the 1936 Berlin Olympics, as this would symbolically bond Germany and Japan. Tokyo was awarded the Olympics the day prior to the opening of the Games in Berlin in 1936.

The Japanese government strategically planned to hold the Grand International Exposition in 1940 as well to highlight "the great achievements of Emperor Jimmu and his illustrious successors who have ruled over Japan in an unbroken line for 26 centuries."[11] The Olympics and the expo were major international events that, if held in 1940, would have served to legitimate the Japanese empire. Promotional material for the events showcased both Japan's tradition and its modernity.

It was hoped that visitors to Japan would see how Japan had rapidly modernized during the Meiji era (1868–1912) and kept abreast of advances in the West, "making up for the delay in modern sciences which has been one of the inevitable sacrifices required by centuries of detachment from other peoples." But visitors were assured that "the indigenous culture and traditions, imbued with the Spirit of Japan, were not forgotten."[12]

> In Japan have converged during the past centuries the main currents of civilization. The country has been a crucible into which have gone all kinds of cultural elements, and its people have lived in the hope that from the crucible will emerge a new fount of civilization that they may call with pride their own. A writer of the West once asserted the modern development of Japan is a miracle the like of which has never been known. The story of this miracle, if that be the proper term for it, is to be told to the world in the Grand International Exposition of Japan in Tokyo and Yokohama in 1940.[13]

It was argued that the main venue of the exposition, Tokyo, "for all of its modernity . . . the city retains much of the charm of Old Japan side by side with its efficient appurtenances of Western nature. The longer you stay and look into the life of the city, the more visible becomes this dual unity."[14] In this way, the exposition was meant to achieve a fusion of both

Eastern and Western civilizations, showcase the development of industry, and promote world "peace."[15]

It was planned to hold the exposition at two venues: reclaimed land on the waterfront of Tokyo Bay around Tsukishima and at Yokohama. The combined sites offered an area of approximately thirty-three square kilometers (816 acres).[16] Most of the pavilions would be at the Tokyo site, with pavilions for oceanography and fisheries and an aquarium at the port of Yokohama. The exposition would open on March 15, 1940, and run until August 30, 1940.[17] A central feature of the exposition would be the Commemoration Hall to the Founding of the Nation, which would showcase images, relics, tableaux, and other displays showing the historical development of the Japanese empire.[18] The Commemoration Hall had the impossible task of needing to be "a permanent structure to be preserved as a piece of Japanese architecture that will embody the essential elements of the styles and techniques of all the distinguished architectural periods of the nation."[19] A design competition for the Commemoration Hall was held in 1937, and it provides us with an example of how architecture was mobilized for political purposes.[20] The winning entry (see figs. 5.2 and 5.3), by Takanashi Shōju, consisted of a building reminiscent of shrine architecture with a tower on top—reflecting the increasingly nationalistic atmosphere in the late 1930s and the role of the state Shintō religion in helping to promote it.[21]

Other pavilions were planned as well. A Hall of Science was proposed that would show not only the development of science but, importantly, how it had been applied to industry. A section was to be devoted to Japanese inventions. The Hall of Art and Education would show how Japan was "the Home of Oriental Art" and "a new Home of Western Art." Part of the hall would be allocated to showing "that the welfare and happiness of mankind depend on healthy development of the spiritual aspects of life."[22]

A building was to be allocated to architecture that would serve as an experimental workshop for the construction industry. It was hoped that modern science could help the Japanese solve building problems through new materials, construction methods, and interior decoration. Examples of Asian architecture would be juxtaposed with Western architecture for comparison, encouraging visitors to see for themselves how an East-West fusion of knowledge could potentially provide a way into the future.[23]

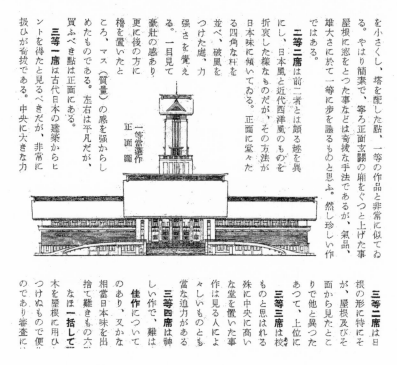

正面圖

一等當選作

三等一席

二等二席

三等二席

三等三席

三等四席

を小さくし、塔を配した點、一等の作品と非常に似てゐる。やはり簡素で、等々正面支關の廂をぐつと上げた事屋根に窓をとつた事などは奇抜な手法であるが、氣品、雄大さに於て一等に歩を讓るものと思ふ。然し珍しい作ではある。

二等二席は前二者とは頗る趣を異にし、日本風と近代西洋風のものを折衷した樣なものだが、その方法が日本味に傾いてゐる。正面に堂々たる四角な杆を並べ、破風をつけた庭、力強さを覺える。一目見て豪壯の感あり。更に後の方に櫓を置いたところ、マス（質量）の感を強からしめたものである。左右は平凡だが、買ふべき點は正面にある。

三等一席は古代日本の建築からヒントを得たと見るべきだが、非常に力扱ひが奇抜である。中央に大きな力

三等二席は日根の形に特にそ屋根及びそ面から見たとこりで他と異つたあつて、上位に

三等三席は校ものと思はれる殊に中央に高いな堂を置いた事作は見る人によ々しいものとも當な迫力がある

三等四席は神しい作で、難は佳作についてのあり、又かな相當日本味を出捨て難きもの六なほ一括して木を屋根に用ひつけぬもので便のであり審査に

Fig. 5.2 Takanashi's design for the Commemoration Hall. *Banpaku*, Dec. 1937, 5.

A building was to have been erected for the tourism industry.[24] Japanese displays at world's fairs in Paris, San Francisco, and New York suggest that monumental, touristic photomurals produced by the design studio Nippon Kōbō are likely to have been a feature. Even after the outbreak of hostilities with China, tourism remained important to Japan. Japan had to publicize its modernity, and tourism was used to shape Japanese and international perceptions of Japan right up to the bombing of Pearl Harbor.[25]

All up, it was estimated that the exposition would cost ¥44.5 million with another ¥15 million for related public works. To cover some of the costs, it was decided to sell admission tickets in advance. Some went on sale in 1938, but war had already broken out with China in 1937 and because of the worsening political situation and shortages in building materials, especially steel, Japan had little option but to withdraw from hosting the Olympic Games and holding the exposition.[26] The Japanese government officially suspended the exposition on July 15, 1938.[27]

肇國記念館模型完成す

丸ビルより三十尺を高いといふ萬國博のテーマ ビルデイング「肇國記念館」の模型が完成されたこの記念館は工費約三百萬圓、樣式は古代日本建築、鐵骨木造四階建、總坪數三千百二十一坪餘、地表から正面中央棟上端迄の高さ實に百二十八尺、即ち百尺の丸ビルを拔くこと約卅尺の宏壯なる建築である。尚内部には特別貴賓室、總裁室を始め正面奧には定席一千七百四十二名牧容のステージ附大講演室あり、左右の翼は繪畫室となり、肇國以來の歷史的事實を一堂に集めることになつてゐる。中央ホールは、壯嚴優美な日本趣味を基調とし、講演室は光線、音響に細心の注意を拂つた近代建築の精華である

Fig. 5.3 Model of the Commemoration Hall, without tower. *Banpaku*, Apr. 1938, 1.

Fig. 5.4 Expo Administrative Building, Tsukishima, Tokyo. *Banpaku*, Nov. 1938, 4–5.

In the meantime, an administrative building (see fig. 5.4) had been constructed on the expo site in Tsukishima, Tokyo, in a section that had been set aside for "Japanese-style" buildings. Completed in September 1937, it is evidence of how architects increasingly were required to convey a sense of national identity in their buildings. There was no grand tower as envisaged in the plan for the Commemoration Hall. Rather, the structure was crowned by a distinctly East Asian–style roof. This was perhaps appropriate given its eventual use by the Imperial Japanese Army. An article in the expo bulletin *Banpaku* (November 1938) announced that the building had been converted to a care facility for sick and wounded soldiers who had returned from the conflict in China.[28]

Technology and the Greater East Asia Co-Prosperity Sphere

In November 1938, Prime Minister Konoe Fumimaro was able to talk of a "new order in East Asia." The historian Daqing Yang has recently argued that the Japanese development of non-loaded cable technology gave the Japanese confidence "as the leader of an Asia that would be independent of Western interests."[29] He has demonstrated how new communications technologies provided both the material and ideological means for Japan to expand its empire and become more self-reliant. As a result, Yang convincingly claims that after the 1930s the Japanese empire became a technologically imagined community.[30] Technology had become a tool of empire in Asia.

The proposed exposition and Olympic Games were not only a celebration but also a stimulus for further technological development. Shin Mizukoshi has written of how the proposed 1940 Olympics encouraged engineers to develop television technology in the hope of telecasting Japan's glory. The Olympics were to be but one of a number of national events from the late 1930s to the 1940s that would feature television technology. These included the Kōha Teishin Tenrankai (Asian Communications Exhibition), Shisō Senden Tenrankai (Ideological Propaganda Exhibition), and Kagayaku Gijutsu Tenrankai (Brilliant Technology Exhibition).[31] In the Grand International Exposition itself, it had been planned to build a Hall of Communications that would be devoted to general communications equipment as well as transportation.[32] Television is likely to have been showcased in that pavilion.[33]

There was now a perception among Japanese that their products were on a par with those of any other country in the world. To account for their success, an English-language promotional guide for the exposition suggested that "there are many reasons, and amongst others we can enumerate abundant labor used to a frugal and simple method of living, manual skill and diligence among the workers, continuous application of the latest technical knowledge, and almost ruthless rationalization, with the elimination of weak members."[34] This Darwinian view of the world and determination would also be seen in Japan's sometimes ruthless expansion of the Greater East Asia Co-Prosperity Sphere. Part of Japan's expansionist policy for Asia was the first "bullet super-express," which was running by 1939. Visitors to Manchukuo could apparently travel the 944 kilometers between Dairen and Harbin in 12.5 hours by riding the "stream-lined, ultra-modern super-express 'Asia.'" Reports describe it as having provided roomy, air-conditioned comfort for passengers, "the last word in speed and comfort." Maximum operating speed was said at the time to have been 140 km/hr.[35] The early attempts to develop television (communications) and the very real expansion of railways (mass transportation) were part of attempts to establish in Japan the "modern apparatus of circulation."[36] Such technologies worked to increasingly globalize the Japanese, introduced new forms of modernity, facilitated social cohesion, and at the same time aided the movement and migration of people from Japan to other parts of the empire. Daqing Yang has argued "that the construction of an ideology of Japan's technological superiority . . . was central to Japan's project of reordering East Asia into a new 'co-prosperity sphere' under its leadership and driving out Western influence."[37] Hiromi Mizuno has explored the key question of how science and engineers in Japan could "embrace both the rationality that denied mythology and the nationalism that promoted mythology."[38] She describes what she calls the rise of "scientific nationalism," something that others have sometimes referred to as techno-nationalism.

Japanese architects, like architects in Germany, Italy, and elsewhere, were also faced by the dilemma of how to reconsider the international outlook of modernism through more local and national values.[39] For the Japanese, Germany and the United States were major sources of modernity. The architect Yamawaki Iwao and his wife, Michiko, were two of only three Japanese who attended the Bauhaus for two years from 1930, after it had moved to Dessau, Germany, and was under the directorship of Mies van der Rohe. The Dessau Bauhaus was dissolved and Yamawaki returned to Tokyo in 1933 and established his own architectural firm. He was involved in the planning and construction of the Japanese pavilion at the 1939 New York World's Fair, working with the design studio Nippon Kōbō.[40]

Japan sought to project an image of both tradition and modernity by having two different exhibits at the fair. The first was a display supported by Japanese businesses in the Hall of Nations. Yamawaki was responsible for the interior design of that section.[41] The second component was the Japanese pavilion that resembled a Shintō shrine and included displays of art and culture within.[42] On the eve of the 2,600th anniversary, the pavilion emphasized Japan's traditions rather than its internationalism and seemed to be more of an exercise in historical revivalism.[43] Credited to the architect Kishida Hideto, a professor at Tokyo University, it appears to have been designed largely by the associate architect Matsui Yasuo, a Japanese architect practicing in New York City at the time.[44] Matsui had worked on many skyscraper buildings in New York, including the Starrett-Lehigh Building on the Hudson River waterfront in the early 1930s. The building had exposed concrete floors and concrete and steel windows, ideas that had been imported from Germany.[45] Matsui, like Yamawaki, was a Japanese interpreter of European modernism, now transplanted to New York City.

We can also turn to graphic design to see how creative Japanese were able to draw on modernist visual techniques to communicate Japan's industrial achievements and also draw attention to its cultural traditions. In the above-mentioned English-language guide to the aborted exposition, Nippon Kōbō was responsible for illustrating the final section on industrial Japan, which included photographs of the Asano Portland Cement Company, the Tokyo Gas Company, Shibaura Electric Works, and the Kanega-fuchi Spinning Company, which the studio photographed. The last page of the guide showed another of their images: a man standing inside a circular generator at the Shibaura Electric Works in Tokyo. While this is evocative of other modernist industrial photographs taken elsewhere in the 1930s, there was a difference. The entire guide was wrapped around with a hand-printed, color woodblock cover with a design that incorporated a landscape scene with Mount Fuji that was evocative of *ukiyo-e* prints of the nineteenth century.

Nippon Kōbō was also responsible for designing the highly illustrated, Western-language, promotional quarterly entitled *NIPPON* (1934–1944), which sought to attract Western tourists to Japan and to promote Japan more generally. The studio's use of photomontages and Bauhaus aesthetics was particularly striking. The magazine showcased how Nippon Kōbō was able to blur "the line between avant-garde art, reportage, advertising practice, and national propaganda."[46]

Nippon Kōbō was cofounded by Natori Yōnosuke, who was born into a wealthy Japanese family. Unable to gain admission to the preparatory school for Keio University in Tokyo, he was instead sent to Germany at the age of eighteen and studied design in Munich. He worked as a photojournalist for newspapers in Munich and Berlin, returning to Japan in 1933. He cofounded Nippon Kōbō with the photographer Kimura Ihei and others who were interested in the Bauhaus concept of Neue Sachlichkeit (New Objectivity), knowledge of which Natori had brought back with him from Germany.[47] What we see in their work is, as Gennifer Weisenfeld puts it, how "reactionary and progressive modernism in publicity and propaganda were cast from the same mold."[48]

Nippon Kōbō was hired to design the Japanese displays in both the Japanese pavilion and the Hall of Nations at the New York World's Fair of 1939–1940. It was also responsible for the interior of the Japanese pavilion at the Golden Gate Exposition in San Francisco in 1939. A major feature of these displays were huge photomurals that used Nippon Kōbō photographs that had already appeared in Japanese promotional magazines. Such photomurals were also a feature of the Japanese pavilion at the 1937 World's Fair in Paris.[49]

Natori has been compared to Leni Riefenstahl. Not only did he introduce photojournalism techniques from Weimar Germany but he returned to Germany in 1936 to cover the Berlin Olympics, contributing photographs to *Berliner Illustrirte Zeitung* (Berlin illustrated news), which was no longer Jewish-owned and was in Nazi hands. He helped fill the void left by Jewish photographers and subsequently transmitted Nazi visual propaganda strategies to Japan.[50] Natori's body of work provides strong evidence of how, during the 1930s, experimental artistic practice informed nationalistic modes of expression that "were equally dynamic and innovative in form, yet unquestionably reactionary in content," supporting government initiatives.[51] Indeed, Natori's activities were backed by the Japanese government. He was associated with the Kokusai Bunka Shinkōkai (Society for International Cultural Relations), a semigovernmental organization that had been established in 1934 to promote cultural understanding. It was chaired by the influential politician Konoe Fumimaro.[52] KBS was a forerunner of the Japan Foundation, which even today is involved in supporting displays of Japanese art and culture abroad.

Natori and his partner Erna Mecklenburg organized two Japanese exhibitions for KBS: one in Leipzig in March 1938 and the other in Berlin from May to July 1938. The aim of the Leipzig exhibition, *Japanese Objects of Everyday Use*, was to highlight the excellent quality and design of objects and tools in Japan. Natori arranged for some four hundred examples of Japanese arts and crafts to be sent to Germany. Not long after the exhibition, on November 25, 1938, a German-Japanese cultural agreement was officially approved and Natori effectively became a bridge between the two countries.[53] The agreement was signed on the second anniversary of the 1936 Anti-Comintern Pact. To commemorate the occasion, the *Japan Times and Mail* newspaper devoted a special issue on that day to the relationship between Germany and Japan. While preparations for Japan's Grand International Exposition had already been suspended, both governments planned to sponsor a program of cultural exchange of films, books, cultural delegations, students, and art shows.[54]

The special commemorative issue included an article by the distinguished Japanese painter Yokoyama Taikan, the text of which was based on an address that Yokoyama had given at a reception in honor of a visiting Hitler Youth delegation. Yokoyama knew that the Hitler Youth delegates would later be visiting Japan's ancient capitals of Kyoto and Nara. He stated that

> during the past sixty or seventy years Japan has been busily engaged
> in introducing in an uncritical manner the material sides of Western
> civilization. But in recent years, and especially now, faced as she is with
> an unprecedented national crisis, the Japanese nation is fully awakened
> to the necessity of going back to the real Japan with its *yamatodamashi*
> [Japanese spirit], and rejecting its own superficial imitation of the West,
> particularly in the field of spiritual culture.[55]

The special issue of the newspaper also contained a large advertisement in German for Leni Riefenstahl's film *Olympia*. However, it was not until June 1940 that part one, *Festival of the Nations*, was released and a longer wait, until December 1940, before the Japanese could see part two, *Festival of Beauty*.[56] It was small compensation for the Japanese who had originally hoped to host the Olympics that year.

The Leipzig exhibition that Natori organized as part of this broader cultural German-Japan cultural exchange was born of a desire on the part of Japan to showcase its technological prowess and its origins in Japanese tradition. Technology would become a symbol of Japan's domination of Asia and help justify the vision of a co-prosperity sphere in East Asia— centered, of course, in Japan.[57] This was certainly the aim of books such as Toyosawa Toyoo's *Sōzō no min, Nihon minzoku* (People of creativity, the Japanese people, 1941), which highlighted the excellence and superiority of the Japanese people, their creativity, and their inventiveness. In addition to more traditional fields such as agriculture, weaving, and arts and crafts, he pointed to how the Japanese had made technological advances in batteries and electrical devices, aviation, weaponry, mining, and transportation.

He also noted Japan's major contributions in scholarly fields such as astronomy, medicine, mathematics, and natural history. He concluded that rather than being thought of as imitators, the Japanese should be celebrated for their creativity.

Japanese-American cultural relations were also singled out for special attention. More funds were channeled toward promoting cultural exchange with the United States in the years leading up to the bombing of Pearl Harbor than in any other period.[58] The *Japan Times and Mail* reported on November 29, 1938, that three hundred pieces of Japanese folk craft would soon be exhibited at the New York Folk Arts Center. KBS had asked Yanagi Sōetsu, founder of the Japanese folk craft movement in the late 1920s and 1930s, to select exhibits after a request from the New York institution was made to KBS via the Library of Congress. The article also mentioned that similar exhibits would be sent to the San Francisco World's Fair and the Ethnographical Museum of Paris in 1939.[59] Japanese craft traditions were becoming the focus of world attention.

The Quest for Japaneseness in Architecture

A key figure who helped promote the importance of traditional Japanese architectural design was the German architect Bruno Taut. In 1933, the same year that Natori left Germany and Hitler came to power, Taut was invited to Japan by the Nippon International Architecture Association. He stayed for three years. On October 30, 1935, at the Peers' Club, KBS invited Taut to give a lecture on Japanese architecture that was translated and published in English as *Fundamentals of Japanese Architecture* (1936) and distributed throughout the world. He praised Ise Shrine, the Kyoto Imperial Palace, and the Katsura Imperial Villa (also known as the Katsura Detached Palace). These imperial structures appealed to Taut in terms of the modernist principles of functionality and rationality. He saw them as authentic, whereas structures associated with the Shogun, such as the Tōshōgū Shrine dedicated to Tokugawa Ieyasu at Nikkō, were deemed to be kitsch. Taut's praise for Ise Shrine and the imperial palaces would have a lasting impact and was interpreted as a victory of Japaneseness over the modern.[60] Taut saw Japan as attractive to foreign architects not because of "the overloaded bizarrerie of Nikkō. Of Japan they had an idealized conception of cleanliness, clarity, simplicity, cheerfulness and faithfulness to the materials of nature, and for the greater part they still retain that conception."[61] Taut effectively argued that traditional Japanese architecture shared characteristics with European modernism.[62] By doing so, he created a lineage for modern architectural practice in traditional Japanese architecture.

Many Japanese believed that Japan combined the best of East and West. This cultural synthesis was considered as surpassing Western modernism. Even back in the Meiji Restoration of 1868 there were calls in the opening decree to return to the events of antiquity and Emperor Jimmu's state foundation, the mythical time before Japan had been contaminated by Buddhism and Chinese civilization.[63] There were thus parallels between the dilemma in which the Japanese found themselves in the 1930s and that earlier time when Japan sought to be more "Japanese" in the 1860s. Fortunately, Taut's views of Japanese architecture accorded with and supported the prevalent nationalistic attitudes of the 1930s.

For modernist architects working in Japan who were resisting nationalist tendencies, Taut's arguments served as a justification for continuing to work in a modernist vein.[64] Taut was not alone in making such arguments. In 1938 the Czech architect Antonin Raymond pronounced, "An architect working in Japan has the advantage of seeing materialized before him, in Japanese architecture and civilization, fundamental principles, the rediscovery of which is the goal of modern architecture."[65] Yet despite such discursive strategies, there was increasing government pressure on Japanese architects in the 1930s to adopt a visibly Japanese style. The answer was the so-called Imperial Crown style, which essentially meant topping a modern concrete, steel-frame building with a traditional, pitched roof. To gain some sense of this response to modernism, we can look to the Gunjin Kaikan (Soldiers' Assembly Hall, 1934)—now known as Kudan Kaikan—in Tokyo, adjacent to the Yasukuni Shrine.[66] In the late 1930s the shortage of building materials and the cancelation of the 1940 International Exposition meant that few plans for major buildings, even those with a Japanese flavor, were realized. However, we can point to Watanabe Jin's design for the main gallery building of the Tokyo Imperial Household Museum, what is now known as the Tokyo National Museum, as a major, late example of an Imperial Crown–style building. Construction of the building began in 1932 and was completed in 1938.[67]

Both buildings were the result of winning designs in architectural competitions. The rules for the Gunjin Kaikan competition stated that "the appearance of the building should express the attributes of magnificence and grandeur while possessing a quintessentially Japanese sense of refinement."[68] Those for the Tokyo Imperial Household Museum specified that designs should be in an "Oriental style based on 'Japanese taste' with a need to preserve harmony between architectural style and the content of the museum."[69]

In 1942 a design competition for a Greater East Asian Co-Prosperity Memorial Hall was held to create a monument to Japan's dreams and ambitions of a new Asian order. The winning entry by the modernist architect Tange Kenzō was what David Stewart describes as an "ultrafascist project," inspired not only by Shintō shrine architecture (especially Ise Shrine) but also by Michelangelo's scheme for the reconstruction of Capitoline Hill and piazza, and Bernini's Saint Peter's Piazza in Rome.[70] It was thus a hybrid design using modern materials such as reinforced concrete. Jonathan M. Reynolds refers to it as "one of the most explicit examples of the deployment of Ise's architectural forms as a symbol of Japanese imperialism during this period."[71] To be located at the foot of sacred Mount Fuji, the memorial hall sought to reflect the East-West fusion that had long been promoted as Japan's answer to how to overcome an overreliance on Western-inspired modernity.[72]

In 1942 Tange responded to a questionnaire about architecture for the Greater East Asia Co-Prosperity Sphere with the statement, "We must ignore both Anglo-American culture and the pre-existing cultures of the Southeast Asian races. To admire Angkor Wat is the mark of an amateur. We should start out with an unshakable conviction in the tradition and the future of the Japanese races. Architects were given the task of creating a new Japanese architectural style in order to contribute to the supreme and inevitable project of the foundation of the Greater East Asia Co-Prosperity Sphere."[73]

That year, the Thai government proposed building a Japan-Thailand cultural center in Bangkok. A competition was held the following year and Tange again won, this time with a proposal that was partly inspired by the Imperial Palace in Kyoto.[74] This reflects how Japanese architects were under government pressure to look to historical examples of Japanese architecture that were associated with imperial sites. And Taut had vouched for their importance and "authenticity."

Overcoming Modernity

We have seen how architects accommodated rising nationalism, but how did intellectuals negotiate the terrain? In July 1942 an important symposium on "Overcoming Modernity" was held in Tokyo, sponsored by the Japanese Council on Intellectual Cooperation. It was modeled after a symposium that was held under the auspices of the League of Nations' Institut International de Coopération Intellectuelle in Europe and South America.

That symposium met nine times between 1932 and 1938, and was chaired by Paul Valéry.[75] It examined the universality of European culture, whether there was such a thing as a European spirit, what role intellectuals and economic needs would play in shaping it, and whether the European would be the "new man" that was emerging in various places at that time.[76] Johan Huizinga was a key participant, giving a lecture on "The Future of the European Spirit" in October 1933. The lecture was followed by a discussion with prominent intellectuals including Valéry, Aldous Huxley, Julien Benda, and others.[77] Valéry's thoughts on the European spirit were published as part of an interview in *Les Nouvelles Littéraires* in 1935.[78] He noted how the European spirit was perhaps on the eve of a disturbance and that "Today every nation tends to isolate itself, to become an island or a closed territory, self-sufficient and living on its own resources. This cannot fail to affect and endanger the intellectual unity of Europe."[79]

Huizinga had also spoken in Brussels earlier that year, on March 8, 1935, and the address formed the basis of his book *In the Shadow of To-Morrow: A Diagnosis of the Spiritual Distemper of Our Time* (1936). Huizinga laments in that book, "How naïve the glad and confident hope of a century ago, that the advance of science and the general extension of education assured the progressive perfection of society, seems to us to-day! Who can still seriously believe that the translation of scientific triumphs into still more marvellous technical achievements is enough to save civilization, or that the eradication of illiteracy means the end of barbarism!"[80] Some of the proceedings were published in Japanese as *Seishin no shōrai: Yōroppa seishin no shōrai* (The future of the spirit: the future of the European spirit, 1936) and *Gendai jin no kensetsu* (The formation of the modern man, 1937).

We can thus view the Japanese symposium as part of international debate over the character of progress.

The Japanese scholars and writers attending the "Overcoming Modernity" symposium gathered to debate how to reconcile Japanese spirit and Western science, that is, how to overcome a largely Western-inspired modernity. For them, the prospect of a modern, non-Western science was difficult to contemplate. The sole scientist to participate was the physicist Kikuchi Seishi, but he remained largely silent. And no architects were in attendance.[81] By this time the previously held idea that the Japanese were able to maintain East Asian values and at the same time acquire Western science and technology came to be seen as problematic, as were simplistic portrayals of East versus West, given Japan's alliance with the Axis nations of Germany and Italy. Nevertheless, some believed that the war itself was a manifestation of the tensions between East and West. Identification of modernity with the West meant that Asian nations seeking to modernize had to become less "Asian" and more "Western." During the war, there were calls to "overcome" this modernity—hence the title of the symposium.

The proceedings of the "Overcoming Modernity" symposium appeared in the September and October issues of the journal *Bungakkai* (Literary world). Richard Calichman has edited and translated essays from the symposium and the transcripts of two roundtable discussions.[82] In an essay entitled "The Heart of Imperial Loyalty" by the novelist Hayashi Fusao, who had been imprisoned for his left-wing activities before and during the war, we see the tensions between East and West played out in the narrative of his life story. He related how the image of the Meiji emperor was on display in the homes and businesses throughout his hometown of Ōita and how over the years the image of the emperor faded away as Japan became prosperous and young Japanese lost sight of their roots and traditions that their parents had held so dear. Hayashi regretted that "the gods had been lost."[83] A preoccupation with learning English and neglect of ancient Japanese texts made young Japanese seemingly less Japanese.

The symposium considered what course of action to take. The philosopher of science Shimomura Toratarō lamented what a fraught exercise they had set themselves given that "modernity is us, and the overcoming of modernity is the overcoming of ourselves."[84] He asked whether it was really possible to negate modernity and argued that it was dishonest to merely focus on the negative aspects of modernity. The film critic Tsumura Hideo addressed the question "What Is to Be Destroyed?" by denigrating American cinema for opposing the modernist spirit with frivolous jazz and wild dance styles. He praised the way German documentary films such as Leni Riefenstahl's *Olympia, Part One: Festival of the Nations* and *Triumph of the Will* surpassed or overcame the modern spirit.[85] He looked to German and Italian cinema as constituting the way forward for film.

In this fascinating way, Japanese intellectuals dealt with the contradictions and tensions posed by the international political situation. The literary and music critic Kawakami Tetsutarō tried to explain away the lack of consensus among the scholars who were gathered by saying, "Our Japanese blood that had previously been the true driving force behind our intellectual activity was now in conflict with our Europeanized intellects, with which it had been so awkwardly systematized."[86] Differences of opinion among the scholars were to be expected given their differing fields of specialization.

The first day of the roundtable discussion was devoted to a discussion of Western modernity and in the second day the focus shifted to the question of Japanese modernity. The literary critic Kobayashi Hideo stated that it was impossible to talk of Japanese literature since the Meiji period without referring to modern Western literature due to the latter's strong influence. He acknowledged that "we must somehow discover Japanese principles, and yet it is quite difficult to achieve this."[87] He continued, "Although we live in the modern age and speak of overcoming modernity, it is clear that the great men of all ages discovered their purpose in life in trying to overcome their own times."[88] Kobayashi later concluded that "we must recognize that modernity is not something that can simply be replaced because of its faults. Modern man can triumph over modernity only through modernity.'[89] Despite the problematic nature of the way in which the symposium had been conceived, it nevertheless gave the scholars an opportunity to reflect on Western influence in their respective fields and look to the future.

For the literary critic Kamei Katsuichirō, in overcoming modernity the Japanese needed to reintegrate themselves with the gods, with the spirit of the Japanese people. Indeed, some participants saw Americanism, the invasion of Japan by American culture, as the result of mass-production strategies developed in the United States. These were seen as undermining Japanese values. As the United States was a relatively young nation, its cultural traditions and morals were seen as shallow and lacking in philosophical depth. The human spirit was considered the creator of technology and separate from manufactured things.[90]

In the final analysis, however, none of the participants proposed abandoning Japan's industrial power. Rather, they saw the human spirit, the Japanese spirit, as separate from technology introduced from the West. The machine was very much the servant rather than the master of human beings.[91] The discussions gave expression to the dilemma felt by many Japanese—a loss of identity as a result of Japan's encounter with the modern and a desire to recover it. In architecture, this took the form of an attempt to create a sense of Japaneseness, even in modern structures that were largely steel and concrete.

Reactionary Modernism?

Was Japan fascist? Emperor Hirohito occupied a role that can be compared to Hitler and Mussolini. Calls to overcome Western modernity were made in the name of a nation and culture united behind the figure of the emperor. The organizers of the Grand International Exposition and its promoters appealed to the ideological domain and used techniques of the mass media to promote an event that would support the idea of an unbroken imperial line dating back to 660 BCE and help justify the expansion of the Japanese empire in its quest for self-sufficiency.[92] We arguably see the suggestion of a reactionary modernism in terms of how Japanese sought to resolve the tensions between tradition and modernity by characterizing Japan's industrial development in terms of Western technology and Eastern spirit, what Marilyn Ivy calls "a fundamental armature of the fascist fantasy in Japan."[93]

It can be argued that there was a fascist culture in Japan with the mythology associated with the emperor system supporting it by offering the idea of a timeless community bound by blood and culture, at a time of upheaval. And it was through propaganda and indoctrination in which the mass media and celebrations for the 2,600th anniversary were a feature. The ideological focus on the emperor drew attention away from the modernizing impulses of the Japanese empire. The Japanese did not reject the benefits of modern technology but sought its roots in their own past, pointing to achievements in science and technology throughout the centuries, obscuring the origins of the largely Western modernity that they had been embracing since the late nineteenth century.

While the management of aesthetics was not so thoroughgoing as in Germany and Italy in terms of architecture, design, and spectacle, the 1940 exposition (and Olympics) shows us the contours of what seems to suggest ambitions that echo those in the Axis nations, local inflections on Japanese modernity that would have to await Japan's defeat before they could be implemented.[94]

Postwar Japanese Modernism

After Japan's defeat the architect Tange Kenzō sought to distance himself from the nationalistic designs that he had created during the war, but he and other architects sought to somehow make their modernist work more distinctly Japanese. Rather than explicitly hark back to historical forms of Japanese architecture such as in Shintō shrines, Tange sought to link Japanese modernism to abstract aesthetic values.[95] Despite the rhetoric, we nevertheless see in projects such as the Hiroshima Peace Memorial Museum, which he designed, a resemblance to the eighth-century Shōsōin imperial storehouse at Tōdaiji temple in the ancient capital of Nara, and columns with proportions reminiscent of Katsura Detached Palace.[96] In a way, this had been unfinished business for Tange and other architects who had been seeking to shape a "Japanese" modernism that drew on Japan's culture and traditions during the war. In the 1960s, Tange made this explicit with his coauthored book *Ise: Nihon kenchiku no genkei* (1962), which was translated and published in English a few years later as *Ise: Prototype of Japanese Architecture* (1965). Tange and his colleagues thus became heirs to Japan's long architectural legacy and key exponents of Japanese modernism.[97]

Before, during, and after the war, we see in Japan solutions for the ill effects and limitations of modern rationalism being formulated in terms of aesthetics, for that is where the essence of Japanese culture (and the nation) was thought to lie. Touristic images at world's fairs of Japan as a place of beauty were more than advertisements to attract visitors; they were a romantic, reactionary attempt to recover that which had been lost, to trans-

輯編局報情
ンセ十・號五十八百二第日八十月八

寫眞
週報

職場は戰場だ

Fig. 5.5 Cover of Japanese wartime magazine showing pilot. *Shashin shūhō* [Photographic weekly], Aug. 18, 1943. Collection: Morris Low.

port visitors to "an alternative world order outside the modern" where reason was replaced by beauty.[98] The ultimate, transcendental image of Japanese identity came to be seen in the figure of the kamikaze pilot whose death was elevated to a thing of beauty, like a cherry blossom, rather than the tragic result of politics and war (see fig. 5.5). The young Japanese pilot represented the Japanese spirit. Although perhaps physically diminutive when compared to Americans, it was his superior spirit and Japanese blood, as a "son" of the emperor, that would make all the difference. Yumiko Iida calls this "aesthetic absolutism."[99]

In plans for the 1940 Grand International Exposition and other projects, Japan turned to the past in order to move beyond the universal framework offered by Western-dominated modernism. The hybridization and historicism that emerged was seen as offering a superior modernism, an alternative modernism, one that would (ironically) flourish decades later in the form of "postmodernism" throughout the world.

▄▄▄▄▄▄▄▄▄▄▄▄▄▄

SIX

EUR

Mussolini's Appian Way
to Modernity

*"Italians, you must ensure that the glories of the
past are surpassed by the glories of the future"*
—*Benito Mussolini*
inscription over the entrance to the Mostra Augustea della Romanità

IN 1935, Benito Mussolini's Fascist government launched the planning for one of the most extravagant world's fairs that never was. The Esposizione Universale di Roma, EUR, was to open in 1942—therefore commonly known as E42—to mark twenty years of Fascism and Mussolini's revival of the Roman Empire.[1] After being delayed by the start of World War II, the fair came to a full stop in 1943 when, as an art historian wryly put it, "Mussolini was otherwise engaged." Though E42 never came to pass, half-forgotten blueprints of the fair preserved in the state archives in Rome provide vivid testimony to the grandiose visions of its Fascist planners.[2]

Beneath the surface of E42 were strong internal tensions between the glories of ancient Rome, which Mussolini famously attempted to recapture, and his Fascist revolution, which he vowed would eclipse even the Caesars. As signaled by the fair's announced theme, "Yesterday, Today, and Tomorrow," the effort to merge past, present, and future was a core dynamic of E42.[3] But, it was a dynamic fraught with complications and contradictions that went to the heart of Fascist culture.[4]

Central to the fair's agenda was a patriotic vision of the future based on progress in science and technology. Like the contemporary science fiction writer H. G. Wells, Mussolini aimed to "invent tomorrow"—although in his case it was a distinctly Fascist vision of tomorrow. Fascist ideologues believed that, with its unique blend of art, technology, and science, Italy was poised to lead the world into the future. And though it had not done much so far in the twentieth century (too bad its great atomic physicist Enrico Fermi had fled with his wife, Laura, to America), it still had an unbeatable team in the likes of Leonardo da Vinci, Galileo, and Marconi.

Yet even as he staked his claim to the future, Mussolini was reinventing himself as the second coming of Caesar Augustus, Rome's first emperor. Underlying his vision of the new Italy and of the exposition was a Fascist myth of national rebirth, which the historian Joshua Arthurs identifies with *romanità*, loosely translated as "the spirit of ancient Rome" or Romanness. An identifying aspect of Italian modernity, it refers to a nostalgic nationalism that flourished in the interwar years.[5] From a more theoretical perspective, the political theorist Roger Griffin dubs it the myth of palingenetic Fascism, distinguishing it from what he calls generic Fascism. Commonly associated with the mystical and metaphysical realms, *palingenesis* derives from the Greek word for "born again." It has a deep history in philosophy, theology, politics, and biology. Griffin points out a paradoxical connection between this backward-looking ideology and forward-gazing modernity.[6] However branded, *romanità* and *futurismo* were the ingredients for a distinctive Italian modernism manifested by Fascism in general and E42 in particular. Griffin further argues that Fascist modernity was tightly bound with notions of scientific and technological progress, an interconnection reinforced by the plans for E42.

Nothing symbolized this incongruous convergence of antiquity with a future wrapped in modern science and technology more brilliantly than the Fair's iconic symbol, the "colossal arch," variously referred to as Arco monumentale or Arco dell'impero (Arch of Empire, see fig. 6.1). Although extensively designed and engineered on paper, it was never built. It was in-

Fig. 6.1 Arco dell'Impero. Courtesy of Anna Maria Bozzola Insolera.

tended to embody the latest that modern science, engineering, and design had to offer. With a silhouette strikingly similar to Eero Saarinen's Gateway Arch in Saint Louis that would be erected two decades later (so similar, in fact, that the architect of the Arco monumentale, Adalberto Libera, threatened to sue the Finnish-American Saarinen for allegedly stealing his design), the imperial arch opened the door to a future based on a faith in scientific and technological progress.[7] Yet just as forcibly it summoned a past associated with the most ancient and recognizable of Roman structures, the triumphal arch.[8] To the Fascist faithful, this blending of past and future was no contradiction; rather, it stood for the marriage of modern science and engineering with the muscular authority and technological prowess of their ancient Roman forebears.

While singularly powerful, the Arch of Empire was only the most spectacular of numerous structures planned for E42 that signified such an amalgam of past and future. What follows is a reading of the arch and other representative artifacts in E42, from exhibitions and museums to monuments, buildings, and city plans. Such artifacts, viewed as an ensemble, encode the paradoxes of Fascist modernism.

Constructing the Fair and Its Meanings

Once the Rome World's Fair received official approval from the Bureau International des Expositions in 1936, planning commenced in earnest. E42 was originally intended to open in 1941, in line with the schedule for official international expositions and in time to celebrate five years of the New Roman Empire, dating from Italy's conquest of Ethiopia in 1936. After falling behind schedule, as world's fairs habitually do, the opening was pushed back to 1942, now marking twenty years of Il Duce's March on Rome and his Fascist revolution.

E42 was to be a showpiece and culminating moment of Fascist culture. Unlike most other world's fairs, it was designed from the start to be permanent.[9] Along with art, architecture, and design, exhibitions enjoyed privileged status within the regime's cultural policies. Wielding aesthetics as a political weapon, the Fascists viewed cultural productions not as ends in themselves—art for art's sake—but, rather, as instruments of state. "Action" was the Fascist watchword, and artists and architects were called upon to lend their active support to the revolution.[10]

Mussolini and Fascist cultural bureaucrats regarded exhibitions as a uniquely powerful tool for swaying mass opinion. During the 1930s a series of expositions propagandized Fascist policies and programs. Among the highly politicized displays, many of them mounted in the Circus Maximus, were the National Exhibition of Summer Camps and Assistance to

Fig. 6.2 Mussolini being shown model of E42. *L'Illustrazione Italiana*, 1942.

Children (1937), the National Exhibition of Textiles (1937–1938), and the Mostra autarchia del minerale italiano (1932–1939; the autarkic exhibit of Italian minerals or ores).[11] Crudely propagandistic, they were showpieces of avant-garde design served up in the brutalist prose of a manifesto.[12] The Mostra della rivoluzione fascista (1932–1938; Exhibition of the Fascist Revolution), to be considered later in this chapter, epitomized this style.

All exhibit roads eventually led to E42, a fair like no other, so its planners believed. Mussolini's name and authority loomed over the planning process; a typical photograph shows him looking imperiously upon a large-scale white model of E42, which is unfortunately lost. As we will see, he used the fair to channel the spirit of Augustus. Billed not merely as international but "universal," the Esposizione was both celebration and justification of his imperial ambitions. It was designed to "bring the world to Rome," writes historian Spiro Kostoff.[13] Rome's mayor Giuseppe Bottai, an influential Fascist instrumental in launching E42, noted that "all the nations of the world" were invited to participate, not just the thirty or so regular participants in world exhibitions.[14] All countries would witness Italy's resurgence. It was the "ultimate hubristic chapter of [Mussolini's] extraordinary career," which would prove that he not only rivaled but had surpassed the Caesars, opines Kostoff.[15]

E42 ranks as one of the most nationalistic expositions of its era, becoming in effect a proxy battleground between Italian Fascism and the world. Yet its subtitle, "Olympiad of Civilization," carried the pretext of internationalism. Even as he invaded Africa, Mussolini marketed the fair as a plea for nonviolent confrontation, a vigorous "gesture of peace, an affirmation of an Augustinian *pax romana.*"[16] Like the Olympics, E42 was to represent national competition within a spirit of internationalism. Mussolini, like Hitler, placed great propaganda value in athletic competition. During the 1938 Tour de France, for instance, he called on all Italians to rally in support of their champion rider, Gino Bartali. Mussolini hailed his eventual victory as a triumph of Fascism and a resurgent Italy. The internationalist mask fell completely away when Mussolini made it clear that E42 would outshine all world's fairs past and future. To highlight the superiority over other national pavilions, Bottai even recommended concentrating Italy's displays in the most prominent plot of the fairgrounds.[17]

Il Duce on Stage

E42 provided the theatrical Duce with a stage for his role as latter-day Caesar.[18] But impressing the world was not enough; to stay in power he aimed to instill awe and fear in the Italian masses. Diane Ghirardo demonstrates how fairs, parades, and mass rallies served his strategy of "surveillance and spectacle."[19] Mussolini reveled in being seen and acting the part, yet he watched the people even as they watched him. Massive images of his head "loomed over and symbolically surveyed the street," an intrusive reminder to Italians of who was in charge. The fair served as a vehicle of mass persuasion, a soft-power hammer used to assert control without resorting to naked coercion.

On stage, Mussolini cast himself as a dual persona. One face was a man of knowledge, seeking public legitimacy by identifying with Italian cultural achievements and deepest traditions, from Roman antiquity to Renaissance Florence. The other face was a man of action, Roman Caesar–cum–Renaissance condottiere—a revolutionary figure "all[ying] himself . . . with the most forward-looking modernizing impulses," indeed as a man of the future.[20]

The Push and Pull of Modernity

Mussolini's contradictory mix of forward- and backward-looking ideology was hardly unique for the era. Whether or not "palingenetic Fascism" was distinctively Italian, propagandizing the past in the cause of modernization was a familiar dictatorial trope. Hitler infamously conjured up Germany's Teutonic origins as he marched the country into a Nazi future. As Don Peretz points out, Kemal Ataturk espoused a modernization strategy of replacing Islamic religion and culture with a restored ancient Turkish past. Reza Shah made a similar bid to recover pre-Islamic glory for an Iran he attempted to modernize through a revival of the myths of the Persian Empire.[21]

Mussolini's appeal to an essential Italianness or Romanness was another instance of what Jeffrey Herf has called "reactionary modernism."[22] Within Mussolinian promodernism were visible seeds of antimodernism: the anti-intellectualism of Fascist exhibitions and aesthetics, a doctrine of ruralism and antiurbanism, the Duce's self-mythologizing as Fascist Superman, and a reactionary longing for a "new man," a Fascist construct purporting to transform the ordinary Italian into a heroic figure embodying manly virtues akin to Mussolini's, but within a strict regime of obedience.[23]

Mussolini's contrasting positions reflected deep socio-psychological needs within destitute, war-torn Italy. As Victoria de Grazia points out, Italians between the wars were profoundly ambivalent about modernity.[24] "Return to tradition" was one of the era's watchwords. For many Italians, modernity subverted traditional values, its democratizing and liberalizing tendencies threatening to upset age-old ways. Stereotyped by many Italians (as well as by Europeans in general during that era) as an alienating, technology-worshipping mass culture, America symbolized modernity's perils. Despite their admiration for many things American—from Hollywood films, popular music, and middle-class living standards to the country's attraction to legions of emigrating *paesani*—ordinary Italians fretted about the implications of "Americanism." For leaders of the regime, it represented a threat to their authoritarian rule. As the anti-Fascist novelist Cesare Pavese put it, the regime tolerated Americanism through "clenched teeth."[25] Even as Mussolini claimed a new day for Italy, he appealed to the nation's classical heritage as a brake on modernity and as a means of forging a unified identity in an impoverished nation torn by class, labor, cultural, and political conflict. This relentless push and pull of modernity generated strong unresolved tensions within the plans for the World's Fair of Rome.

EUR as Stage and Symbol

Although E42 never materialized, it nevertheless left a permanent legacy in the exposition district known as EUR—Mussolini's Fascist City of the Future.[26] Today a lively Roman suburb, EUR provides an overarching context for parsing the exposition's complex ideologies. First and foremost, it was a monument to totalitarian planning: Mussolini saw the fairgrounds as the model for the renovation of historic Rome in his image. After considering various sites within the precincts of the capital, he eventually decided to build E42 from scratch on an undeveloped 420-acre parcel about five miles south of the city center. Under the general direction of Marcello Piacentini, Italy's preeminent architect and a Mussolini favorite, EUR's planners imagined their Fascist utopia, its gleaming white buildings set temple-like in a verdant realm of gardens, parks, and lakes.[27] The new suburb represented a step toward Mussolini's long-cherished dream of extending Rome to the sea, even reestablishing the Mediterranean as the revived empire's mare nostrum.[28]

The concept of EUR partook of Mussolini's program of *città di fondazione*, "foundation cities," planned towns designed as the new basis for his revolutionary Fascist state. In consolidating his power, Mussolini gave highest priority to the reform of cities. Casting a suspicious eye on the metropolis, he regarded it as breeding ground for social unrest and subversion. To realign the city with Fascist ideology, he launched a massive program of urban reconstruction. The program involved a variety of strategies: renovating sectors of existing cities, creating new urban districts linked to, but outside, the boundaries of existing cities (as EUR), and building cities and towns de novo, mostly in rural areas. The latter approach of marrying town and country drew inspiration from the garden city movement of the early twentieth century.

In the new-built category were "technology cities" such as Torviscosa —dubbed the City of Cellulose—and the Città dell'Aeronautica, two of the thirteen foundation cities that were ultimately built. These new Fascist towns, loosely anticipating today's high-tech ex-urban hubs, demonstrated Mussolini's commitment to science-based technology. Most were erected in the Italian countryside, modified garden cities centered on large industrial projects. Architecturally, Il Duce's new towns typically blended modernity and tradition; factories, city halls, and other government buildings were built in the style of Fascist modernism while housing and other domestic structures tended to embody vernacular styles.[29]

These new country-based industrial towns were designed to generate the technology and materials necessary for autarky, the Italian quest for economic self-sufficiency. They also played an important symbolic role, with their opening ceremonies feeding Mussolini's propaganda machine. At the official launch of Torviscosa in 1938, for example, two towering pylons were erected temporarily to perform the symbolic function of the traditional triumphal arch. Adorned with Mussolini's Latinized title, Dux, they reinforced his self-identification with ancient Rome. Mussolini and his retinue strode triumphantly between the twin structures to the enthusiastic applause of officials, workers, and townspeople. The event became fodder for propagandistic newsreels, in which the pylons were projected against a backdrop of the chemical towers of Torviscosa's synthetic fiber plant. In scenes reminiscent of Riefenstahl's *Triumph of the Will*, the factory's chemical towers were dressed out in the form of the Fascist axe. Michelangelo Antonioni, newly embarked on his storied directorial career, was later commissioned to produce a promotional documentary about the new town's creation.[30] Mussolini's ambitious new town program showcased his ability to commandeer an urban space and reshape it in the Fascist image.[31] In essence another foundation city, but one with government as its core industry, EUR demonstrated these talents on a grand scale. It, too, straddled past and future, traditional and modern.

Soon after planning for E42 began, Vittorio Cini, the senator appointed by Mussolini as president of the quasi-independent company that was developing the fair, wrote: "The Exposition of Rome will try to create the definitive style of our era: that of the year XX of the Fascist Era, the style of 'E42.' It will obey criteria of grandeur and monumentality. The meaning of Rome, which is synonymous with eternal and universal, will prevail—it is to be hoped—in the inspiration and execution of constructions destined to endure."[32] What his words meant was far from clear; concepts of monumentality, authentic "Fascist style," and universalism, as well as notions of modern and traditional, were still up for grabs. The ensuing architectural and ideological debate pitted a well-established tradition of Italian modernism led by Giuseppe Pagano, architect and editor of the magazine *Casabella*, against the monumental Fascist classicism associated with Piacentini, who claimed his vision was the only authentic expression of *Italianismo*. Although it is not possible here to detail this critical debate about the future of Italian architecture, suffice it to say that Piacentini prevailed.[33] EUR is known today for its unrivaled concentration of monumental buildings in the Fascist-*moderne* style, a white city in limestone, tuff, and travertine marble.

Mixing modern geometric forms with such classical motifs as colonnaded arches, Mussolini's updated utopia represented a futuristic, scientized version of classical Rome. While a portion of EUR was built in the period 1937–1943, the bulk of it did not see completion until the postwar era. But even these later-built structures still used the initial Fascist template.[34] Among major prewar structures was the Palazzo della civiltà del lavoro (Palace of working culture), nicknamed the Colosseo Quadrato (Square Colosseum), for its six tiers of arches and geometric rendering of Rome's most famous monument. Built between 1938 and 1943, it was designed by Giovanni Guerrini, Ernesto Bruno La Padula, and Mario Romano. At the top of the high-rise structure is this evocative patriotic inscription: "A People of Poets, Artists, Saints, Thinkers, Scientists, Sailors, and Explorers." Since the Arch of Empire never materialized, the Square Colosseum has emerged as the iconic symbol of EUR. It epitomizes Fascist rationalism, a futuristic take on Roman classicism that became the district's signature style.[35]

Adjacent to it is Gaetano Minucci's Palazzo degli uffici dell'Esposizione Universale di Roma, the headquarters for EUR planning and development. The building is surrounded by heroic statuary, fountains, and mosaics, some crafted by noted futurists.[36] At the entrance to the complex is a popular tourist attraction, a marble slab that traces the saga of Rome from Romulus and Remus to Il Duce on horseback delivering his traditional Fascist salute.[37] Another impressive prewar building, designed by Luigi Figini, Gino Pollini, and Mario de Renzi, was the headquarters of the

Fig. 6.3 The Square Colosseum. Rigamonti Archive, Department of Image Collections, National Gallery of Art Library, Washington, DC.

Italian army. Today it houses Italy's central state archives, which hold the records of E42 (see fig. 6.4).[38] Besides these early government structures, several new museums were also underway. Organized axially along wide boulevards, the city plan of EUR echoes the stark geometry of its buildings, exhibiting an almost mathematical discipline of power and authority. Somewhat softening this martial precision are the district's many green spaces, its gardens, fountains, lakes, and other engineered water features.

Arising in the 1920s and lasting well into the 1950s, rationalism was an enduring architectural tradition that, at least in theory, stood for progressive ideals of logic and scientific rationality. Though eschewing traditional Italian decorative motifs, rationalism introduced the ornamental vocabulary of Fascist modern.[39] Among its notable practitioners were Marcello Piacentini, Adalberto Libera, Giuseppe Terragni, Gino Pollini, and Luigi Figini. The movement was traditionally identified with modern technology, both as practice and symbol. Rationalist builders pioneered the use of innovative construction methods and materials such as steel and reinforced concrete. They also employed modern building and transportation techniques that allowed building rapidly and on an unprecedented scale,

Fig. 6.4 Architect's rendering of the building that now houses the Central State Archives, reminiscent of de Chirico. Courtesy of Anna Maria Bozzola Insolera.

techniques that allowed Mussolini to realize his most extravagant architectural dreams.[40]

Rationalism harked back to the Italian futurists, proponents of the avant-garde art movement launched in 1919 by the poet F. T. Marinetti. Enthusiastically embracing Fascism and its culture of violence, they identified themselves with modern science and Machine Age technologies, most famously with aviation.[41] Futurism remained a vital intellectual current within Fascism. "The EUR has a de Chirico-like perspective," remarks architectural observer Vincent Scully, referring to the iconic futurist artist Giorgio de Chirico (1888–1978). The Greek-born Italian artist was known for his "metaphysical" paintings featuring eerie images of neoclassical buildings on deserted city squares, often with steam locomotives, factory smokestacks, and other industrial markers depicted in the background. What Mussolini aimed to build, writes Scully, was "a haunting image of Italian tradition. And the Fascists were sometimes able to do just that, as in the EUR, site of the Esposizione Universale di Roma."[42]

Arco dell'impero, Gateway to Modernity

The signature structure being planned for the Esposizione Universale di Roma was the unbuilt Arch of Empire, the Arco dell'impero (also known as *Arcone* dell'impero because of its gigantic size). It was intended above all to impress. Images of this towering sculpture, designed by Adalberto Libera, one of Italy's premier rationalist architects, in concert with engineer Vincenzo di Berardino, were regularly featured in illustrated magazines, posters, and other promotional material for the fair.[43] At 240 meters high and 600 meters across the base, the tapered semicircular arch brilliantly captured the spirit and ideology of E42. The scientifically designed structure was to stand at the main entrance to the exposition, straddling the road between central Rome and EUR. It was to be a futuristic marvel of engineering art. Exceeding the Eiffel Tower in grandeur, if not in height, it represented a thrust into the scientific and technological future. Yet, at the same time, it evoked the most traditional of Roman forms, the triumphal arch. In a 1938 letter to Vittorio Cini, Pier Luigi Nervi, Italy's most famous engineer-architect, along with an enthusiastic cadre of fellow engineers, hoped to persuade Cini and Il Duce to approve the colossal arch as E42's chief symbol: "[the Arch of Empire] will be of vast proportions, genial and of the highest technical Italian Art and besides being an absolute novelty it will constitute the grandest conception in the field of Technique and Mechanical Art. . . . The Duce and the highest Authorities of the Regime have warmly approved the idea borne out of the minds of four Italian Civil Engineers."[44]

It was just the sort of "high-concept" that Mussolini and his loyal followers found irresistible. In short order Il Duce signed his approval, declaring: "Arco = E42 e vice-versa."[45] As Duce, he had passed through countless triumphal arches, temporarily erected for ceremonial occasions, in "the manner of the victorious generals of antiquity or of Renaissance nobles and clerics."[46] This gateway was special, however. The Arch of Empire was to be a permanent monument to Mussolini and his new Roman Empire. Mirroring the ancient arches of Constantine and Titus in the Roman Forum, it was both homage and modernist answer to antiquity. Its advocates guaranteed that it would outshine all other European monuments:

> Having dimensions of 600 meters in width at the base and of 240 meters in height . . . the proposed project, whose dimensions are exceptionally noteworthy and such as to [easily compete with such contemporary structures as] the Cathedral of Notre Dame in Paris, St. Stephen's in Vienna, St. Peter's in Rome, Saint Paul's in London, and the Duomo in Pisa with its Campanile, will immediately be seen by those who are directing the competition as the sought-for element that will possess the characteristics and world-class grandeur so as to truly become a curiosity equal [to] if not greater than the Eiffel Tower.[47]

The call for proposals for the Arch fired the imaginations of inventors, engineers, and architects throughout Italy. They strove to showcase the capabilities of Italian civil engineering, especially the mastery of innovative construction materials like concrete, aluminum, and chromium steel. The E42 archive overflows with engineering drawings, blueprints, feasibility studies, and data tables for the proposed structure. Technical documentation includes geological soundings of the site by Nervi, mathematical analyses of structural stability under extreme conditions, technical comparisons between materials such as masonry, aluminum, and steel, and calculations as to which material was most conducive to autarky.[48]

In reviewing proposals, E42's jurors wanted to be assured of the soundness of the enormous and innovative structure. One solution to the stability problem was the "reverse arch," proposed by Dr. Engr. Ettore Fenderl. He designed a matching underground arch that would firmly anchor the aboveground portion of the structure. He argued that the full-circle construction would prove more solid than a half-circle, guaranteeing that the structure would not fall down. It had the added advantage, he said, of doubling the volume of the arch, which would allow visitors to make a complete circuit of the monument.[49] Although Fenderl's idea was not adopted, such proposals convinced E42's organizers, at least initially, that Italian engineering was up to the task, and that the arch could and should be built. Two options for materials were finally proposed for the arch, chromium steel—a rust-resistant stainless steel much in vogue in the art deco period—and unreinforced concrete. The latter version was the one ultimately incorporated in the official plan of 1938, although many of the publicity images still showed a lustrous stainless-steel polish.

Once the critical structural issues had been addressed, other important questions remained: what would happen on the inside of the structure, and what purposes would it serve? Proposals were nothing if not ambitious. A. Ferretti, for example, proposed an elaborate internal transportation system, though his offer to make a model was not accepted.[50] It was eventually decided that four internal cog railways would transport visitors around the structure. At the apex would be a restaurant with a bar and an amusement plaza, featuring a spectacular parachute drop. Each leg of the arch would stand on a stout pedestal, housing heavy machinery, electrical plants, warehousing, and maintenance personnel. While anchoring the structure, the pedestals would also contribute to its overall aesthetic, their bulk accentuating by contrast the lissome lines of the arch's apex.

The engineers who proposed the arch also pointed to its dual symbolic function of honoring both ancient Rome and modern Italian engineering: "The construction of the colossal Arch, besides glorifying the Roman origins of architecture, provides the most extensive mechanical applications, of curiosity, attraction, light, and lighting effects, as would be otherwise difficult to obtain for our event."[51] The planners' call for lighting effects sparked special interest. In the spirit of the "Olympiad of civilization" and Mussolini's peace rhetoric, Antonio Voltaggio suggested that the arch should stand as a symbol of peace and light beaming from Rome. He proposed flooding it in diffuse light at night or, even more magically, outlining it in multi-colored neon, "like a great rainbow emanating from Rome"—a harbinger of peace for a tortured world.[52] Others took this idea further, suggesting that the Arch could provide a futuristic advertising surface, a gigantic billboard on which to project light shows promoting corporate products. It would be the ultimate emblem of Mussolini's corporatist city, exemplifying the Fascist ideal of government-corporate partnership.

The notion of using the Arch as advertising space spoke to the core goals of the fair. Publicity was central to the concept of E42. In fact, provision was made for a major pavilion of "Pubblicità," showcasing Fascist innovation in advertising and "propaganda," a term that did not carry the pejorative connotations it does today. The pavilion would show the "function of advertising as instrument of defense and of conquest . . . [and] of Italian commerce."[53] Publicity and propaganda were regarded as vital instruments of power, even signs of the modern state. The 1920s and 1930s witnessed massive growth in the advertising and public relations industries, spurred by the rise of radio, TV, and other electronic mass media. Accordingly, besides featuring traditional newspapers and periodicals, plans for the Italian *pubblicità* pavilion included movies, radio, posters, expositions, and exhibits, electrically illuminated signs, and store windows.

The Arco monumentale was originally conceived as a propaganda tool, both for the fair and the regime, but it was not to be. Despite the enthusiasm for building it, it posed huge financial and technical problems. Finally conceding that the technical challenges were causing unacceptable delays, E42's planners jettisoned the magnificent structure.[54] The arch would survive only as a vivid memory of what might have been.

Marking EUR's Museum District: The Marconi Obelisk

Just as the Paris world's fair of 1837 eventuated in the permanent display of science in the Palais de la découverte, E42 was expected to leave an array of permanent museums as well as an entire new museum district. Marking the entrance to this future museum quarter was an imposing white obelisk dedicated to the Italian radio pioneer Guglielmo Marconi, an inventor who stirred special pride in the hearts of Mussolini and all Italians. Although as a structure the obelisk was less spectacular than the Arco dell'impero, it had the advantage of actually being built. It was equally symbolic of Il Duce's dual commitment to modernization and to antiquity. It was designed under Mussolini's 1937 commission by Arturo Dazzi, who had previously sculpted a monument to Marconi for the Italian pavilion at the 1939 New York world's fair. It was eventually completed in 1960, in time for the Rome summer Olympics. The obelisk stands today on EUR's Piazza Marconi, the district's main square. Reliefs on one of the obelisk's marble faces celebrate the Italian radio pioneer's life and his revolutionary contribution to communications. An adjacent face features themes of Christianity, the Fascist dictator's quid pro quo to the Vatican for the pope's acceptance of his government.[55]

As a monument honoring Marconi, the obelisk's profile was singularly appropriate, its vertical thrust clearly evoking a radio antenna and technological modernity. In the 1920s and 1930s, electronic communications, spearheaded by the invention of long-distance wireless radio, were considered leading-edge technology. Radio had emerged as the world's most powerful mass medium, bringing about massive societal transformation. Mussolini immediately seized upon its power to reach the masses. Like Hitler (and, for that matter, FDR, though for different purposes), he proved a master of the medium and used it to expand and secure his dictatorial authority. Standing at the center of EUR, envisioned by Il Duce as his new seat of power, the obelisk could be read symbolically as a radio tower beaming his orders to his far-flung empire.[56]

Fig. 6.5 Marconi Obelisk on Piazza Marconi. Rigamonti Archive, Department of Image Collections, National Gallery of Art Library, Washington, DC.

At the same time, the obelisk ranks among the most ancient of technological emblems, one of the great, not fully understood technical feats of early antiquity. In the first master plan for EUR, the 1,700-year-old Ethiopian obelisk that Italian troops looted from Axum in 1937 during their African campaign was to be re-erected in EUR to represent the launching of the new Empire. The final version of the master plan, however, relocated the Axum obelisk to a site within historic Rome. The Marconi obelisk was then built to replace it in EUR. Designed in the Ethiopian style, the newly built obelisk clearly referred to the Axum original (which was repatriated to Ethiopia only in 2005), thus adding empire to its rich vocabulary of associations. This single structure became a kaleidoscope of meaning. Even as it memorialized Marconi's pioneering role in communications, radio, and technological modernity, it referenced African antiquity, the Holy See, the empire, and dictatorial power. The rationalist-style museums built on the Piazza Marconi reinforced the monument's ideologies. Still another monument to Marconi—a museum of communications—was originally in the offing for EUR but was never built. Other major museums on the

Piazza Marconi, however, were realized before the war. The Museum of Folk Art and Traditions and the Museum of Prehistoric Ethnography remain imposing reminders of Fascist ambitions of cultural hegemony.

Exhibition of the Fascist Revolution

Another major museum intended as a permanent installation in EUR was the Mostra della rivoluzione fascista (Exhibition of the Fascist revolution). Its special connection with the Duce and his Fascist regime is well documented in the scholarly literature.[57] To emphasize here, however, is its blending of modernist and traditional motifs.

Conceived by Alberto Alfieri, director of the Institute of Fascist Culture, the exhibition captured Mussolini's personal interest, for obvious reasons. It embodied the same blend of modernity and tradition that stamped the plans for E42.[58] The exhibit opened in 1932 in the nineteenth-century-era Palazzo delle esposizioni. Adalberto Libera was commissioned to redesign the Palazzo's original façade. Known as an archrationalist, Libera prided himself on the use of innovative materials, the latest construction technologies, and principles of industrial design. The museum's catalog called his redesigned façade a hymn to modernity. At the same time, the building incorporated the motif of the triumphal arch that Libera aspired to take to new heights in E42's Arco monumentale. Reinforcing the modernist stylistics of the project was the popular illustrator and Fascist artist Mario Sironi, who collaborated with Giuseppe Terragni in laying out the building.[59] Updated in the late 1930s to reflect the anti-Semitic turn in Fascist ideology with the promulgation of the anti-Jewish Racial Laws of 1938, the new—and final—version of the Mostra became vulgarly nationalistic. Major additions included displays warning of the Communist and Jewish menace. But the racial laws proved highly unpopular among ordinary Italians and visitation to the exhibit fell off sharply with this last revision.[60] Plans for converting the exhibit into a permanent Museum of the Fascist Revolution collapsed after the Duce's fall in the summer of 1943.

E42, which was being developed around the time of the revision of the Mostra della Rivoluzione Fascista, promoted the same nationalist goals and likely would have resembled it in design and graphic style. However, official anti-Semitism does not play a visible role in the plans for E42, at least as revealed so far in the archival record. Though a rising factor in Fascist culture, such rhetoric may well have been muted or suppressed for the occasion of the fair. Even fervent Fascists must have been concerned about how anti-Semitic themes would have played out on an international stage.

The Museum of Roman Civilization

Mussolini planned to integrate into E42's museum complex a blockbuster archaeological exhibit on Roman civilization, the Mostra archeologica. Its inclusion would seal his identification with Augustus. First held in 1911 at the Baths of Diocletian to commemorate the fiftieth anniversary of Italian unification, the display was moved to a temporary venue in Rome in 1929. Seizing upon what he saw as a major propaganda opportunity, Mussolini incorporated the exhibit in his grand plan to glorify Augustus, a plan that had already commenced with his renovation of the emperor's mausoleum. In 1937 Mussolini ordered the exhibition moved to the Palazzo delle esposizioni, the museum that had previously presented the Mostra della rivoluzione fascista. Rebranded the Mostra augustea della romanità (Augustan Exhibit of Romanness), the archeological display was repurposed to celebrate the two thousandth birthday of Augustus.[61] Yet, even as he dedicated the display to Augustus, Mussolini was determined to upstage him. Still looming today over the exhibition's entrance is this exhortation by the self-anointed modern Caesar: "Italians, you must ensure that the glories of the past are surpassed by the glories of the future."

The Mostra augustea della romanità was an educational enterprise. Its didactic displays consisted primarily of reproductions—casts of statues, inscriptions, reliefs and parts of buildings, models of monuments and architectural complexes, home furnishings, and models relating to early Roman science and engineering. When the exhibit moved to the Palazzo, it was significantly expanded with the addition of more replicas.

After the Augustan exhibit ended its run at the Palazzo delle esposizioni, Mussolini ordered it moved to a new dedicated building in E42. Just as the never-held Japanese International Exposition celebrated the 2,600th anniversary of the ascension of the legendary Emperor Jimmu (see chapter 5), the Fascist fair would now officially mark the two thousandth anniversary of the birth of the first Roman Emperor. Begun in 1939 under a commission by Fiat, the new building was designed by Ascheri,

Fig. 6.6 Model of Roman Vinea, Portable shelter used in assault on fortresses. Roya Marefat.

Bernardini, and Pascoletti in EUR's high Fascist style as a celebration of the Italian Duce. Interrupted by the onset of war, it was completed with architectural revisions in 1952. Fiat eventually donated the building to EUR, where it reopened in 1955 as the present Museo della civiltà romana (Museum of Roman Civilization). Now joined by an astronomical museum and planetarium, it is part of a museum complex near the Piazza Marconi. Consisting of twin connected buildings, the museum is a monument to Fascist architectural pomposity. A grand portico of arcaded marble columns joins the two wings and spacious interior halls soar to impressive sky-lit heights.

Major exhibit halls of the museum are devoted to Roman civil and military engineering.[62] Large-scale dioramas showcase examples of Roman construction, including colorful models of aqueducts, bridges, buildings, the Appian Way, and other roads. Roman military technology is presented as the epitome of engineering and military prowess. Among the attractions are realistic models of vineas, catapults, battering rams, and other siege engines.

Striking a similar martial theme are the museum's casts of the bas-reliefs scrolled around Trajan's column, erected in the Roman Forum in 113

CE. Made by the Vatican in 1861 for Napoleon III, the casts occupy one of the museum's largest halls. The reliefs present a narrative of the many exploits of the emperor, focusing on his legendary battles that saved Roman civilization from invading barbarians. A public relations ploy on behalf of Trajan, they now served as historical validation for Il Duce's imperial adventures, echoing the martial rhetoric of his propaganda campaigns.[63]

For the leader who had just announced the new Roman Empire, the exhibit's central attraction was made to order. An enormous white scale-model, an intricate jigsaw puzzle of some 125 pieces, presents Imperial Rome in the age of Constantine, when the Empire attained its peak expansion. By highlighting not the Rome of Augustus but that of Constantine the Great, the emperor who legitimized Christianity, the model, like the Marconi obelisk, was another bow to the Vatican.[64] The model encompasses the area within the Aurelian Walls, which enclosed the Seven Hills of Rome. It features exquisite replicas of the Colosseum, Circus Maximus, and other charismatic monuments. Encapsulating the landmarks of imperial Rome within the walls of a Fascist museum was another not-so-subtle way of supporting Mussolini's claim on past glories.

The Museo della civiltà romana also incorporated exhibits of early Roman science and medicine, including a large-scale model of an astrological globe, a hemispheric sundial, and reproductions of scalpels and other early surgical instruments. Several other areas of the museum, including the hall of musical instruments, present basic scientific principles. But the major address to Italian science and technology was reserved for another exposition venue.

Museo delle Scienze

The Mostra scienza universale, the exhibition of "universal science," was a centerpiece of the Italian world exposition. It, too, was slated to become the basis for a permanent Museum of the Sciences, Museo delle scienze. The science museum presented a comprehensive survey of the history of science. Though the project never got beyond the drawing board, it was heavily planned. "Universal science" was the titular theme, but the focus was actually on Italian scientific genius, in line with Il Duce's nationalistic cultural policies.

Fig. 6.7 Model of Rome in the time of Constantine. Roya Marefat.

Mussolini staked the future of his Fascist regime on Italian primacy in science and technology. As Denis Mack Smith has shown, Mussolini made it clear that scientists and inventors could not afford to be politically neutral.[65] He expected them to join arms with artists and other intellectuals in advancing the Fascist cause. The Museum of the Sciences in E42 offers a case study in how Mussolini bent scientists, inventors, and scholars to his ideology. We will see that while their plans for the fair's science and technology exhibits were not overtly propagandistic, they clearly advanced an agenda of cultural nationalism.

An opinion piece in an ephemeral Fascist journal, *L'Organizzazione scientifica del lavoro*, vigorously endorsed the project, arguing that a new national museum of Italian scientific ingenuity was long overdue. Dubbing it Museo italiano del lavoro (Italian Museum of Work), the writer invoked the model of the Deutsches Museum in Munich, long recognized as the international standard for science museums.[66] An emblem of "Autarchia" on the journal's back cover asserted the principle of Italian scientific independence.

The displays would show the world that, after an inexcusably long period of "lethargy," Italy was finally prepared to resume its traditional role as global scientific leader. The Regime's propagandists made much of this Fascist scientific "reawakening"—a second Italian *rinascimento*.[67]

Presiding over the exhibition's planning committees was Francesco Giordani, a chemist from the University of Naples who would later become a key figure in establishing Euratom, the European Atomic Energy Community. Sabato Visco, the director of the Institute for Physiology at the University of Rome, was his vice-president. Soliciting exhibition and architectural proposals from the country's scientists, engineers, architects, designers, and scholars, they received a swift and enthusiastic response about how to display Italy's claim on the scientific future.

But, first, to jump-start the Mostra scienza universale, the organizers went back to Leonardo. Seeing a guaranteed hit in a Leonardo exhibit currently on display in Milan, they planned to bring it down to the Rome world's fair after it finished its run in Milan. Leonardo's star power would be a major public draw and add instant legitimacy to the Italian science pavilion, as well as to the fair in general. *Ingegni e congegni* (Engineers and devices), another little-known Fascist engineering journal, reviewed the Leonardo show in glowing terms.[68] The exhibition, wrote the reviewer, featured Leonardo as both artist and inventor and included displays of his manuscripts and models of his inventions. For Italian patriots, the Leonardo exhibition would lend the aura of a genuine superstar to E42 and buttress Italy's "grand battle for autarky." Bringing it to Rome would be a major propaganda coup for the Fascist regime. Sabato Visco himself took the train to Milan to see it for himself, with the plan of identifying elements suitable for presentation in E42.[69]

The Leonardo project received Mussolini's personal endorsement. As a self-styled "man of knowledge," he identified personally with Leonardo and promoted scholarship and exhibitions on the history of science. He did much to revive Leonardo studies, although his intentions were hardly academic. As Galluzzi points out, the Fascists exploited the Renaissance genius for propaganda purposes, perhaps exaggerating his reputation in the process.[70] Others suggested honoring Galileo at the exposition as well, noting that 1942 was the three hundredth anniversary of the legendary experimenter's death.[71] The proposal to honor Galileo represented another approach to salvaging Italy's scientific reputation. As the inventor of the experimental method, it was argued, Galileo could be credibly portrayed as the true father of modern science. The inclusion in E42 of Leonardo and Galileo, the twin pillars of Italian science and invention, would signal to the world that Italy was back as a scientific power.

There was one problem with this scenario, however: the two Italian greats, however ahead of their times they may have been, still lived and worked centuries ago. Save for Marconi and some outstanding work by Italian biologists, the Italian performance in science and technology over the last century was distinctly subpar by world standards. According to Galluzzi, E42's planners worried about this disparity between fantasies of Italian scientific ascendancy and the country's backwater reality.[72]

Various rationalizations for Italy's relatively low scientific standing were offered. Some apologists argued, with a certain degree of irony, that nationalistic biases of other nations had blinded them, perhaps willfully, to Italy's scientific greatness. Others contended that the world failed to recognize the importance of fundamental discovery, Italy's forte, as compared to the sort of routine research churned out in vast quantities elsewhere. Regardless of the reality, Mussolini called the tune, demanding that Italian scientists find a way to demonstrate Italy's priority in discovery and invention.[73] Galileo invented the scientific method, so the argument went, laying the foundations for all subsequent research; electrical experimenters such as Ampère and Faraday were only building upon the prior discoveries of Galvani and Volta, and so on. According to one Giorgio Abetti: "Italy is in an eminently favorable position regarding its contributions to science in the civilized nations, because in fact Italy nurtured the origins of the experimental method, an event of fundamental importance that led to great developments through the work of Galileo and his disciples in showing new and unsuspected avenues of research." In the area of invention, Il Duce insisted on Italian priority for the dynamo, the telephone, the typewriter and other major technological discoveries. In his proposal for the exhibition, M. M. Apollon argued that it must include a special section on invention that would prove "the profundity of Italian scientific intuition."[74]

Mussolini began to lobby internationally on behalf of Italy's scientific and technological revival, emphasizing the theme of Italian originality. Fascism's directive policies, he maintained, were leading to an Italian resurgence in science and technology, as well as in other intellectual areas. He contrasted Fascist science policies with the free investigation encouraged by Italy's pre-Fascist regimes, which he disparaged as a failure. Undirected science and scientists, he asserted, gave rise to contradictory notions like Einstein's perplexing theories, a "Jewish fraud lacking in originality."[75]

Exhibit plans for the new museum included most of the physical and biological sciences. Technology and engineering were also well represented, but chiefly as applications of scientific discovery—one consequence of scientists taking the lead in the planning process for E42.[76] Giordani laid out a straightforward and hierarchical, if overly literal, floor-by-floor scheme for the Museo delle scienze. At the museum's basement level were exhibits on geology and the "underground sciences," the ground floor documented scientific origins (Leonardo, etc.), the first floor featured Galileo charting the path to the modern physical sciences, and medicine and biology were explored on the second floor. The plan called for a second building devoted to the astronomical sciences, connected by a bridge to the main museum.

Displays of the history of technology and engineering were largely relegated to the wings or to annex buildings, where they were presented as applications of the basic sciences. Critics of this approach took strong exception to this privileging of science at the expense of technology. Fascist ideologues in particular balked at the implied hierarchy, demanding that technology be given a more central role in the museum. Engineering, they pointed out, was a traditional Italian strength going back to antiquity. Technology exhibits in the nearby Museum of Roman Civilization were designed to show that modern engineers stood on the shoulders of Roman forebears. In technology and engineering, the path to modernity was surely by the Appian Way, according to these critics, who also maintained that technology and engineering were subjects much more accessible to the public than basic science.

Architectural designs proposed for the Museum of the Sciences followed the conventions of classic Fascist modernism, in effect cloaking modern science in the trappings of Roman antiquity. For the interior of the building, the firm of Brusa, Cancelotti, Monturori, and Scalpelli, winner of the architectural commission, submitted elaborate plans for a three-story colonnaded structure at a central location of the fair. Interior elevations show a lofty central hall running the length of the building.[77]

A twenty-two-meter-high glass dome covers a central courtyard, while an imposing statue of an unidentified figure stands on the ground level directly beneath. A central grand staircase rises to the four exhibition levels. Interior decoration includes generous use of ornamental brick and stonework, as well as balconies overlooking the main hall. A medallion set beneath the dome honors Romulus and Remus at the breast of the she-wolf.

Plans for decorating the astronomy hall bespeak the grandiosity of the vision. Flowery quotations from Virgil, Ovid, and Homer, among other classical poets, are inscribed over the entrance. One of the vestibules showcases a cast made from the Vatican Museum's Urania, the Greek muse of astronomy.[78] The Museum of the Sciences would present Italian science and engineering in the reflected glory of classical antiquity and myth.[79]

Exhibits in the individual science halls were organized along the spine of a historical timeline, laying out parallel histories of the various scientific specialties. While there is no space here to present those individual narratives, it can be said that, viewed as a whole, the plans are rather conventional and do not make for the most exciting reading. Galluzzi deemed them a distinctly amateur effort, reflecting the general ignorance of the scientific planning committees about the history of science. The committees dithered, never reaching consensus, leaving the whole exhibit program up in the air—the most likely reason for the failure of the museum to materialize.[80] But, to give due credit, even a failed attempt to create what amounted to a university-level course in the history of science and technology may have been a first (and probably a last) for world's fairs, which traditionally feature superficial displays of the wonders of science and technology.

Potentially more interesting were alternative (or complementary) plans for displaying Italian scientific progress. Rather than exhibiting science and engineering in a dedicated museum of science, this alternative called for integrating them within a broad-brush Mostra della civiltà italiana, an exhibition on Italian civilization. This plan, which also failed to come to pass, was to portray scientific and technological developments within a comprehensive historical context that interwove themes from art, architecture, music, religion, literature, theater, the history of exploration, and military history. As in E42 generally, science and technology would be presented as the through-line of Italian history.[81] Granted, many of the proposals struck a note of Italian, if not Fascist, triumphalism, but not as stridently as one might have expected, given the temper of the times.

In fact, one redeeming aspect of the plans for both the Museo delle scienze and the Mostra della civiltà italiana was their relatively restrained propagandistic tone, at least as originally rendered. To be sure, there was the requisite flattery of the Duce, including proposed equestrian statues, and halls of heroic tribute. But they were strictly window-dressing; none was fleshed out in any way, at least in the surviving documents—which goes to show how difficult it can be to control committees of scientists and scholars, even under totalitarian regimes. That said, of course, at least some of the obligatory Fascist embellishments would probably have been added at a later stage by watchful ideologues.

The Museo delle scienze bespoke the value that Mussolini's regime placed on science and engineering. While plans for the building followed classical architectural lines and invoked the names of ancient poets, these embellishments were not atypical of science museums in general. None of this was in fact inconsistent with scientific modernity. To be sure, Mussolini's anti-Semitic laws of 1938 took a heavy toll on Italian science, forcing scientists at, or destined for, the top rank, like Enrico Fermi, Emilio Segrè, and Rita Levi-Montalcini, to go into hiding or to flee the country. Nevertheless, "reactionary modernist" themes on the Italian cultural scene do not appear to have compromised the basic content and representation of science. The proposals submitted for consideration display none of the romanticized or irrational anti-science currents that Gerald Holton and other historians have detected in Nazi Germany among conservative, anti-modern scientists.[82] Fascist faith in a modernity based on scientific and technological progress remained essentially intact.

Ambiguous Legacy

Although the aborted fair was all but forgotten after World War II, it gradually resurfaced in Italian national memory. Two anniversary exhibitions based on the archival documents were held, in 1987 and 2005, and there is a growing scholarly literature on the fair and Fascist culture.[83] Despite its cancellation, E42 left an indelible physical imprint on Rome, both on the city center and on suburban EUR, even as it was completed after the war. The planning process forever altered the overall contours and look of historic Rome. Mussolini was obsessed with rebuilding the capital in his image, the most visible way for him to compete with Augustus. Even before construction of EUR began, Mussolini had ordered a strategic demolition of whole sections of the city, focusing on areas around famous monuments such as the Colosseum and the Forum.[84]

Reimagining Rome both as it once was and as it was to be, he described this project as the "liberation" of the monuments from the slums and the clutter that had engulfed them during the "centuries of decadence" after Augustus, robbing them of their visibility and splendor.[85] Once liberated, Rome's great monuments regained their past glory and stood out like architectural jewels. Mussolini's editing of the capital city involved major rebuilding: removing the detritus of the past freed up space for remaking Rome along the lines of EUR, giving the ancient cityscape a rationalistic Fascist overlay that remains to this day. To Mussolini, monumentality was the essence of the Fascist style. One of his most grandiose dreams went unbuilt: his idea for a gigantic Mussolini Forum, centering on an eighty-meter-high symbol of Fascism, described by Denis Mack Smith as "a half-naked figure of Hercules, one hand holding a truncheon, the other raised in a Roman salute, the face bearing Mussolini's features."[86] The project was abandoned, however, when money ran out.

Diane Ghirardo's observation that "any government architecture has a rhetorical function [that] tells us what the regime wants us to believe about its nature" sheds important light on the form of EUR.[87] Its powerful rhetorical messages not only spoke to the times but also cast a long shadow on the postwar art and cultural scene. In the 1960s, after the district acquired much of its present form, it took on a quasi-mythic role in Cinecittà, the Italian film capital. It is the featured backdrop in a number of iconic postwar Italian films, including masterpieces by such auteurs as Federico Fellini, Michelangelo Antonioni, Bernardo Bertolucci, and Pier Paolo Pasolini.

No longer a Fascist version of a scientifically based paradise, however, it is utopia gone sour. It is transformed into a symbol of postwar sterility, alienation, and angst. In Antonioni's *Eclipse* (1962), EUR is portrayed as a barren, dispiriting cityscape. Becoming virtually a character in itself, it holds the fretful Vittoria, played by Monica Vitti, in its psychic grip. As a symbolic form, the graceful Arch of Empire was replaced by something to quite the opposite effect. An ominous water tower, called Il Fungo for its mushroom-like shape, looms over Vittoria, evoking both H. G. Wells's Martian mechanical invaders and an atomic mushroom cloud, an especially chilling image in the year of the Cuban Missile Crisis. Science has become a dystopian force. With the intimidating mushroom-like structure lurking in the background, the film verges on science fiction and surrealism reminiscent of de Chirico. Was Vittoria fleeing Martians invaders, the atomic bomb, or the sinister shadow of Fascism? Or all at once? It is a curious irony, however, that Il Fungo, completed in 1960, was originally designed to serve a quite different, much more benign function. It was built to protect the city's green spaces, providing reserves of water to irrigate EUR's parks and gardens and to suppress fires that potentially threatened them. Today the former water tower has been repurposed as a popular rooftop restaurant with a panoramic view of the city.

Likewise, in the final scenes of Pier Paolo Pasolini's heavily ideological dark comedy, *The Hawk and the Sparrows* (*Uccellacci e Uccellini*, 1966), the iconic geometric skyline of EUR, from the Square Colosseum at one end to the Cathedral of Saints Peter and Paul at the other, rises ominously on the horizon. In these films and others, EUR's buildings, once emblematic of a Fascist utopia, fit far from comfortably in historic Rome, or in contemporary Italy, for that matter. Clearly, by the decade of the 1960s the signature architectural style adopted by EUR had outworn its welcome.[88]

These powerful films have created an unforgettable, yet arguably one-dimensional memory of EUR, or at least what it was intended to be. To Mussolini and his architects, this was the dawning rationalist utopia. How was it possible for them to think of it that way? Archival records reveal the Fascist ideals behind the putative utopian concept of E42. Its most alluring emblem was the Arch of Empire, a brilliant fusion of the past and the scientific future. It was pure ideology frozen in concrete. Although the arch was never realized, it lives on in myth as part of an imagined tomorrow shaped by an omnipotent techno-science. For good or ill, the myth was partially realized, in exhibitions, museums, monuments, and other stylized edifices that purported to invoke the past in order to win the future for Mussolini's Fascist paradise. The question remains, however: Was going back to the future ever a viable strategy? To paraphrase the fair's tagline, "Yesterday, Today, or Tomorrow?"

For E42's planners, all was done in the name of the dictator. After submitting their proposal for the Arch, Libera, Nervi, and his fellow engineers waited anxiously for word from Il Duce. Then came the hoped-for decision when Mussolini penned "Arco = E42" on their proposal. The new Caesar had spoken. "Let it be written; let it be done" was the tacit mantra of the fair's planners. Yet, despite Il Duce's decrees, the Appian Way to modernity was no magical time machine; it was a tortuous journey. The path was never smooth, nor was the destination ever guaranteed.

SEVEN

REDEFINING CULTURAL CONFLICT

The First Postwar International Exposition

WORLD'S fairs were the flamboyant venues of serious "soft power" struggles. Under a veneer of entertainment, backed by a significant commitment of resources, olympiads of ideologies were taking place. The international expositions of the 1930s provided an opportunity for nations to establish their roles in a new, emerging global order and—equally importantly—to convince their own peoples of the worthiness of the sacrifices they were being asked and would soon be asked to make. Each nation made a claim upon the future based on a synthesis of the mastery of the tools of modernity (especially technology, science, and art) and particular national strengths. For the future Axis powers (Germany, Japan, and Italy) this approach involved an ideology melding modernity and national and racial myths. "Reactionary modernism," to use the term coined by Jeffrey Herf, involved the interweaving of romantic, irrationalist ideas with the spectacular rationalism of modern technology.[1] As we have seen in the case of Japan, industrial development was reimagined as the blending of Western technology with Eastern spirit. Nazi Germany took great pains to appear to its own people and to the world as both avant-garde and tied to its timeless Germanic roots. Italian Fascists, too, saw no contradiction in this blending of reverence for the authority of ancient Rome and the efficacy of modern science and engineering.

For the Soviet Union, it meant dramatic claims to rapid material progress in one country based upon socialism, implying for them and their supporters the inevitable triumph of the Soviet system. The expositions mounted by the left-of-center governments in France and the United States were far more sensitive to domestic politics and to intellectual differences among the expositions' planners. The planners of the 1937 Paris Fair cut across the political spectrum. It was broadly agreed that France's continuing role in the modern world was intellectually to provide a reasoned approach to modern life, and aesthetically to provide a high level of good taste. Diplomatically, the French center yearned to provide a middle way between the foreign regimes of the left and right. The situation was different for the United States. To most Americans, the nature of modernity seemed far less problematic. For the overwhelming majority, technology promised a richer, healthier, and more satisfying future, but the question remained: who was to lead the way, and how? Americans focused on tensions between free enterprise and planning, working out their definition of liberty and justice for all within an economy increasingly dominated by large business enterprises.

The Second World War completely altered the playing field. Germany and Japan were abolished as great powers, and Italy abandoned its imperial aspirations, seemingly forever. France and Great Britain, strength sapped, faced the dissolution of their own vast empires. Two powers, now labeled "superpowers," remained on the playing field: the Soviet Union and the United States of America. The half-century after the war ended in 1945 has subsequently been popularly characterized as the period of the Cold War and the Atomic Age. The wartime alliance between the United States and Joseph Stalin's Soviet Union quickly fell apart and was replaced by economic, diplomatic, cultural, and propaganda warfare, spiced by espionage and the occasional blockade. The American monopoly on atomic weapons was ended by the Soviet Union in 1949 and the so-called nuclear club was soon joined by Great Britain (1952), France (1960), and China

(1964). A major "hot war," the Korean War of 1950–1953, while officially a "police action," saw the United States and China engaged in open warfare. It threatened to plunge the world into a nuclear disaster.

In a world beset by such problems and the urgency of reconstruction of wartime ravages, world's fairs were not matters of major immediate concern. The first postwar "universal and international" exposition was mounted in Brussels, Belgium, in 1958, years after the end of the war. This world's fair marked the continuation of the pattern of ideological struggle established during the 1930s, but this time the battle was between the two major superpowers, with the rest of the nations providing a supporting cast.

The Belgians, naturally, wanted an exposition for reasons of their own. The director of the fair, Baron Georges Moens de Fernig, had been minister of trade in the Paul-Henri Spaak government and responsible for the implementation in Belgium of the Marshall Plan. Under the Baron's direction, the Belgian planners wished to advance the cause of Brussels as the new capital of Europe. In 1957, western European nations established two new communities, the European Economic Community (EEC), with its commissions centered at Brussels, and the European Atomic Energy Community (Euratom), with its administrative headquarters also at Brussels. The theme the commissioners chose for the expo stressed both cooperation and peaceful uses of nuclear energy. Officially, it read: "The Balance of the World for a More Humane World." The subtitles go on to explain: "Technology in the Service of Humankind" and "Human Progress through Technological Progress."[2] A fuller account of the commissioners' motives may be gleaned from the instructions sent to the US planners from the Belgian Information Service. The Belgians wanted to "create a crossroads of the nations" wherein would be worked out "the duty of cooperation" among nations. They particularly emphasized atomic energy, information, and the new media. They asked America to highlight international cooperation and humanism.[3]

The symbol of the fair was the *Atomium*, a 334-foot-high representation of an iron molecule consisting of spheres representing atoms, each of which was almost sixty feet in diameter, large enough for exhibits or restaurants, that were linked by escalators or walkways. Many nations, especially Belgium, France, the United Kingdom, the United States, and the Soviet Union, presented displays of their use of nuclear energy for electri-

Fig. 7.1 Interior, Japanese pavilion. Courtesy of Wouter Hagens.

cal generation. France described its plans to generate over a quarter of their energy requirements within a decade. The United Kingdom provided an exhibit called "Power for Progress" that displayed a scale-model nuclear power station and an exhibit explaining British research in nuclear fusion.[4] The former Axis powers (Germany, Italy, and Japan) all were represented and each assumed a very low profile. In striking contrast to their prewar pavilions, their exhibitions were modest in size, low in silhouette, and sent the message that all three were ready and willing to participate in the family of nations and its desire for a peaceful world.

While the architecture of the exhibitions of Japan and Italy stressed their vernacular styles, Germany's unimposing pavilion was, in Greg Castillo's words, "a spectacle of restraint."[5] Even though it had twice the floor space of the Paris 1937 Deutsches Haus, the 1958 building projected the feeling of a smaller, lighter project. Alain de Botton, in his *Architecture of*

Fig. 7.2 Exterior, Soviet pavilion. Courtesy of Wouter Hagens.

Happiness, notes that whereas in Paris the German pavilion was "ominous, aggressive and defiant," in Brussels the architecture emphasized "horizontality to suggest calm, lightness to imply gentleness and transparency to invoke democracy."[6]

By way of contrast, the Soviet pavilion stressed its scientific and industrial achievements, alongside its cultural attainments. While planning for the exhibition began in 1956, the direction of the presentation took a dramatic turn after 1957 when the Soviet Union launched the world's first artificial satellites, Sputnik I and Sputnik II. Especially at the urging of V. D. Zakharchenko, the editor of the magazine *Technology for Youth*, Soviet achievements in space began to assume an increasingly larger role.[7] The Soviets issued a weekly newspaper at the fair, *The Sputnik*, piously wishing that "all nations should march together on the road to progress." Models of the two Sputniks were prominently displayed alongside a model of a future Soviet "cosmic space platform" powered by the light of a lamp.[8] Marquis Childs of the *Washington Post* noted the "visible national pride in the faces of the crowds that stream in and out of the building" in "this fantastic new achievement of science."[9]

Fig. 7.3 Interior, pavilion of the USSR. Courtesy of Wouter Hagens.

The architecture of the pavilion itself was notable for its radical departure from traditional socialist realist Soviet forms. Designed by a team of architects including Alexander Boretsky, Anatoly Polyakov, Yuri Abramov, and Viktor Dubov, it was often described in the press as "the refrigerator" for its supposed likeness to the household appliance.

Within, the Soviets displayed not only their achievements in space but also their big machines and industrial wares. Howard Taubman, the *New York Times*'s cultural critic, described it as follows:

> The Russians have adopted the hard sell. . . . [They] appear to take loud exuberant pride in their wares. They are trumpeting their products, achievements, people and ideas. . . . The Russians, in short, behave as if they have a clear and definite idea of what they wish to achieve with their pavilion. They have arrived at Brussels confident and optimistic, fresh from such scientific triumphs as the launching of their sputniks. . . .

They are making much ado about their growing industrial power. . . . People go through the Russian pavilion remarking on its impression of size, strength and material well-being.[10]

Taubman relates his discussion with a young Belgian "of good education, middle-class background and democratic conviction." The young man said of the Russians, "They're strong. Maybe you Americans are still too complacent."[11]

It was just this aura of complacency that the United States wished to avoid. The responsibility for Brussels 1958 was placed in the hands of the Department of State, and in January 1957 President Eisenhower named Howard Cullman, who was a tobacco magnate, a director of the New York Port Authority, a director of the Metropolitan Opera, and a well-known Broadway investor, as commissioner general. He was ably reinforced by James Plaut, director of the Boston Institute of Contemporary Art, as his assistant commissioner.[12] The State Department, the US Information Agency, and the commissioners interviewed over fifty Americans prominent in all areas of corporate and public life, including Thomas J. Watson of IBM; Walter Paepcke of the Container Corporation of America; Walt Disney; Edwin Canham, the editor of the *Christian Science Monitor*; Nelson Rockefeller; labor leader Walter Reuther; and historian Arthur Schlesinger Jr. In Schlesinger's view, the exhibition should not shy away from America's problems (well-publicized around the world) but, rather, show an America facing up to its problems and making progress solving them and this resonated with many public intellectuals.[13]

The commissioners turned to the Massachusetts Institute of Technology for expert advice on how to integrate science and technology into the fair. In April 1957 they called together a weekend workshop with the fair planning committee and MIT notables charged with developing concrete ideas for the exhibition. The group's chair was Elting Morrison, and included (among others) the economist and director of MIT's Center for International Studies (CENIS) Max Millikan, Walt W. Rostow, physicists Victor Weisskopf and Martin Deutsch, and political scientist Ithiel de Solla Pool. The Cambridge Study Group, as it was known, shaped a broad consensus that the fair commissioners saw as depicting the American people as a people engaged in constant change, a dynamic people confronting their

Fig. 7.4 United States of America pavilion exterior. Courtesy of Wouter Hagens.

challenges in a creative way.[14] Rostow's Idealism in Action subcommittee formed the idea for what became the *Unfinished Work* exhibit at the fair. Acknowledging that America's racial problems and urban blight were well known to the world, especially after the crisis in September 1957 at Little Rock's Central High School, the exhibit aimed at showing candidly that Americans had problems but were making principled progress in solving them.[15]

The US exhibition, housed in Edward Durrell Stone's pavilion, was plagued by budget problems. The original appropriation was $15 million (compared with the Soviet's estimated $60 million and the Belgian Congo's $10 million) but even this sum, which Cullman considered embarrassing, was cut by $4.5 million early in 1957. Supporters managed by 1958 to restore some small funding to around $12 million.[16]

The exhibition opened on time. Visitors entered and were greeted by a kaleidoscope of "American" images: a Model-T Ford, a jukebox, license plates, the Sunday *New York Times*, cowboy boots, and so forth. The path led to the art exhibit featuring folk, Native American, and contemporary art and sculpture. Along a circular path, visitors encountered "Atoms for Peace," with irradiated foods, a medical radiation unit, and electro-mechanical hands that allowed the visitor to "handle" radioactive materials. The theme of continual industrial revolution was displayed by a card-playing computer that challenged all comers to a game of bridge, and an IBM computer that answered questions of historical dates in ten languages. At President Eisenhower's insistence, the exhibit included a voting machine at which visitors could vote for their favorite US president, musician, and movie star. The winners: Abraham Lincoln, Louis Armstrong, and Kim Novak. A live theater featured Benny Goodman, Harry Belafonte, and the Philadelphia Orchestra. On the second floor over a thousand consumer items featuring top designers were displayed. One of the most popular attractions at the fair was the fashion show running throughout the day that had beautiful live models. Leaving the exhibit at the second floor led the visitor to Walt Disney's Circarama (360-degree) film *America the Beautiful* and to *Unfinished Work*, the controversial exhibit that came out of the MIT meeting. In the latter, visitors were led through three sections. In the first, major social problems facing America were laid out: self-government, equal rights for the nation's black population, and urban blight. The second section, "the people take action," proclaimed that "the doom of the American caste system is in sight." The last section, "hope for the future," displayed photomurals expressing confidence that "democracy's unfinished business" would be completed. Predictably, angry Southern senators and congressmen decried what they saw as an anti-Southern bias. Others worried about washing the nation's dirty linen in public. The outcry worked. The exhibit was removed and replaced by a public health exhibit.[17]

Many Americans expressed disappointment that the US pavilion did not meet the Soviet "hard sell" effectively. The press quoted an Atlanta businessman who said that "the exhibit is a national disgrace," and a Chicago visitor who was dismayed that "nowhere [do you] get a real feeling for what this country is like." The foreign press reaction appears much more favorable. The London *Tribune* is quoted as saying, "The American reply to the Russian steam-roller offensive is of surprising subtlety and

Fig. 7.5 Interior, US pavilion. Courtesy of Wouter Hagens.

intelligence." The Oslo *Aftenpost* stated, "America introduces itself with such charm and pleasant ingenuity, it will doubtless impress Europe a great deal."[18] But Americans themselves were less friendly to the "soft sell." Humorist Art Buchwald wrote, "Not only are we soft selling the United States at Brussels, in reality we've built ourselves a giant tranquilizer pill." But he noted, "We are winning one major propaganda victory at Brussels. While all the other pavilions and exhibits are charging admission to the washrooms, the United State is offering free washrooms! People from all over the exhibition are thronging to the U.S. pavilion, and we have won more friends through our washrooms than any other exhibit in the house."[19]

The American soft sell in 1958 may have been roundly criticized at home, while having a better reception in Europe. But for many reasons

Washington, deeply enmeshed in the Cold War, seemed stunned by the public relations success of the USSR with Sputnik, and embarked on an answer to Sputnik that would be reflected in subsequent world's fairs. Just as the Brussels exposition opened, President Eisenhower's Science Advisory Committee released to the public a promissory note. This little pamphlet, issued from the White House on March 26, 1958, was titled *Introduction to Outer Space*, and described four main reasons to undertake a national space mission. The first and most important was the "curiosity that leads men to try to go where no one has gone before."[20] The promissory note was redeemed in time for the *Apollo 11* exhibit at the Osaka World's Fair in 1970.

NOTES

1. Introduction: World's Fairs, Modernity, and the Demand for Authenticity

1. Anne O'Hare McCormick, "National Exhibitionism at the Paris Fair," *New York Times*, July 24, 1937, 14.

2. On German ambivalence towards "Americanization," see Mary Nolan, *Visions of Modernity: American Business and the Modernization of Germany* (New York: Oxford University Press, 1994), 11–12, 108–30, and "America in the German Imagination," in *Transactions, Transgressions, Transformations: American Culture in Western Europe and Japan*, ed. Heide Fehrenbach and Uta Poiger (New York: Berghahn Books, 2000), 3–25. On French views, see Roxanne Panchasi, *Future Tense: The Culture of Anticipation in France between the Wars* (Ithaca, NY: Cornell University Press, 2009), 117–27, and the contemporary Georges Duhamel, *America the Menace*, trans. Charles Miner Thompson (London: Allen and Unwin, 1931). For a comparative view, see David Ellwood, *The Shock of America: Europe and the Challenge of the Century* (Oxford: Oxford University Press, 2012).

3. Eric Hobsbawm, "Mass-Producing Traditions: Europe 1870–1914," in *The Invention of Tradition*, ed. Eric Hobsbawm and Terence Ranger (Cambridge: Cambridge University Press, 1983), 263–307.

4. Maurice Roche, *Mega-Events and Modernity: Olympics and Expos in the Growth of Global Culture* (London: Routledge, 2000), 34–64.

5. Paul Greenhalgh, *Ephemeral Vistas: The Expositions Universelles, Great Exhibitions and World's Fairs, 1851–1939* (Manchester: Manchester University Press, 1988), 49.

6. Karl Marx and Friedrich Engels, *Basic Writings on Politics and Philosophy*, ed. Lewis Feuer (Garden City, NY: Anchor Books, 1959), 10.

7. Anthony D. Smith, *Nationalism and Modernism: A Critical Survey of Recent Theories of Nations and Nationalism* (London: Routledge, 1998), 117–42, 224.

8. Bruce Strang, "'The Worst of all Worlds': Oil Sanctions and Italy's Invasion of Abyssinia, 1935–1936," *Diplomacy and Statecraft* 19, no. 2 (2008): 210–35.

2. Modernity à la française: The 1937 Paris Exposition

1. Shanny Peer provides a brief, charming "walking tour" of the Exposition, and as of this writing several useful home movies by visitors are available on the Internet: Peer, *France on Display: Peasants, Provincials and Folklore in the 1937 Paris World's Fair* (Albany: SUNY Press, 1998), 42–51; for film tours see the Prelinger Archive, http://archive.org/details/0367_HM_Medicus_Collection _Paris_International_Exposition_1937. On Speer's German pavilion and its context, see Karen Fiss, *Grand Illusion: The Third Reich, the Paris Exposition, and the Cultural Seduction of France* (Chicago: University of Chicago Press, 2010).

2. See for example Paul Dupays, *L'exposition internationale de 1937: Ses créations et ses merveilles* (Paris: H. Didier, 1938).

3. Peer, *France on Display*, 21–51, skillfully outlines the origins of the exposition from its earliest parliamentary proposals in 1929 to its cancellation and rebirth. The political scene is discussed in Madeleine Rébérioux, "L'exposition de 1937 et le contexte politique des années trente," in *Paris 1937: Cinquantenaire de l'exposition internationale des arts et techniques dans la vie moderne*, ed. Bertrand Lemoine (Paris: Institut français d'architecture/Paris Musées, 1987), 26–29. We have benefited greatly from research notes supplied by Prof. Miriam Levin of Case Western Reserve University.

4. Charles R. Day, *Schools and Work: Technical and Vocational Education in France since the Third Republic* (Montreal: McGill-Queen's University Press, 2001), 35–37.

5. Charles R. Day, *Education for the Industrial World* (Cambridge: MIT Press, 1987), 63–64.

6. Roger Dautry, "Culture and Technique," *Harvard Business Review* 12 (1934): 409–12.

7. Paul A. Gagnon, "French Views of the Second American Revolution," *French Historical Studies* 2, no. 4 (1962): 430–49. See also a contemporary survey, William Spoerri, *The Old World and the New: A Synopsis of Current European Views on American Civilization* (Antwerpen: Imprimerie du Centre, 1936); Ellwood, *Shock of America*, 13–214; Seth Armus, *French Anti-Americanism: Critical Moments in a Complex History* (Lanham, MD: Lexington Books, 2007), 1–56; David Strauss, *Menace in the West: The Rise of French Anti-Americanism in Modern Times* (Westport, CT: Greenwood, 1978); Jean-Philippe Mathy, *Extreme-Occi-*

dent: *French Intellectuals and America* (Chicago, Chicago University Press, 1993), 53–103.

8. Duhamel, *America the Menace*, xiii, 213.

9. André Siegfried, *America Comes of Age: A French Analysis* (New York: Harcourt and Brace, 1927), 25–26. Siegfried was elected to the Académie française in 1944.

10. Siegfried, *America Comes of Age*, 180–81.

11. Peer, *France on Display*, 26–27.

12. "'Babbitt' Gets Invitation to Paris 1937 Exposition," *New York Times*, Nov. 9, 1934, 14.

13. *Exposition Paris 1937* 1 (May 1936), rear cover, recto. The magazine was unpaginated for the first month.

14. Paul Léon, "Each to His Kind," *Exposition Paris 1937* 12 (May 1937), 15.

15. Edmond Labbé, *Paris, Exposition Internationale des Arts et Techniques dans la vie moderne (1937), Rapport Général* (Paris: Imprimerie nationale, 1938–1940), i, xi. Translation of the report is made by the authors.

16. Labbé, *Le régionalisme et l'exposition internationale de Paris 1937* (Paris: Imprimerie nationale, 1936), 122; and Levin, unpublished notes provided to the authors.

17. Labbé, *Le régionalisme*, leaf 9 recto.

18. Labbé, *Le régionalisme*, leaf 2 recto–leaf 3 verso.

19. Pascal Ory, "Le Front Populaire et l'Exposition," in Lemoine, *Paris 1937*, 30–35; Peer, *France on Display*, 33–42. See also Gilles Ragot, "Le Corbusier et l'Exposition," in Lemoine, *Paris 1937*, 72–79.

20. Gilles Plum, "Le Palais de la Découverte et le Grand Palais," in Lemoine, *Paris 1937*, 294–99. A more detailed look can be found in Ory, *La Belle Illusion: Culture et politique sous le signe du Front Populaire 1935–1938* (Paris: Plon, 1994), 479–84.

21. Jean Perrin, "The Palace of Discovery," *Exposition Paris 1937* 2 (June 1936): 5–6.

22. Perrin, "The Palace of Discovery," *Exposition Paris 1937* 12 (May 1937): 9.

23. An extensive and laudatory review of the Discovery Palace was provided by California Institute of Technology physicist Jesse W. M. DuMond, "The 'Palace of Discovery' at the Paris Exposition of 1937," *Journal of Applied Physics* 9, no. 5 (1938): 289–94.

24. Perrin, "Palace of Discovery" (June 1936): 5.

25. Perrin, "Palace of Discovery" (May 1937): 6–9.

26. Peer, *France on Display*, 26, 192n19.

27. For an analysis of the mural, see Rosi Huhn, "Art et technique: la lumière," in Lemoine, *Paris 1937*, 400–401.

28. Peer, "Radio," in Lemoine, *Paris 1937*, 242–43.

29. Bertrand Lemoine and Philippe Rivoirard, "Electricité et Lumière," in Lemoine, *Paris 1937,* 222–23.

30. See also Huhn, "Art et technique," 398–99.

31. Gilles Plum, "Chemins de Fer" in Lemoine, *Paris 1937*, 218–19.

32. Andreas Fickers, "Presenting the 'Window on the World' to the World: Competing Narratives of the Presentation of Television at the World's Fairs in Paris (1937) and New York (1939)," *Historical Journal of Film, Radio and Television* 28, no. 3 (2008): 297.

33. Alison J. Murray Levine, *Framing the Nation: Documentary Film in Interwar France* (New York: Continuum, 2010), 89, 95–97.

34. McCormick, "National Exhibitionism," 14.

35. See among other scholarly works Dawn Ades, "Paris 1937: Art and the Power of Nations," in *Art and Power: Europe under the Dictators 1930–45* (London: Thames and Hudson, 1995), 58–62; Fiss, *Grand Illusion*; Danilo Udovicki-Selb, "Facing Hitler's Pavilion: The Uses of Modernity in the Soviet Pavilion at the 1937 Paris International Exhibition," *Journal of Contemporary History* 47, no. 1 (2012): 13–47; and Sarah Wilson, "The Soviet Pavilion in Paris," in *Art of the Soviets*, ed. Matthew Bown and Brandon Taylor (Manchester, UK: Manchester University Press), 106–19.

36. Anne O'Hare McCormick, "Little Paris on Last Fair Day Shows France Still Safe," *New York Times*, Nov. 27, 1937, 16.

37. Udovicki-Selb, "Facing Hitler's Pavilion," 34–35. See also Jean-Louis Cohen, "U.R.S.S." in Lemoine, *Paris 1937*, 184–89.

38. "Indeed, national socialist architecture was above all an art of staging." Dieter Bardzko, "Allemagne," in Lemoine, *Paris 1937*, 137.

39. Bardzko, "Allemagne," 139.

40. Fiss, *Grand Illusion*, 71.

41. Barbara McCloskey, *Artists of World War II* (Westport, CT: Greenwood Press, 2005), 93.

42. Labbé, *Paris, Exposition Internationale*, 9:11.

43. Labbé, *Paris, Exposition Internationale,* 9:11.

44. On the relationship with Terragni's Casa del Fascio and the controversy between innovative and conservative architects, see Richard A. Etlin, *Modernism in Italian Architecture, 1890–1940* (Cambridge: MIT Press, 1991), 477; and McCloskey, *Artists of World War II,* 93.

45. Labbé, *Paris, Exposition Internationale,* 9:12.

46. Labbé, *Paris, Exposition Internationale,* 9:15.

47. McCloskey, *Artists of World War II,* 94.

48. Labbé, *Paris, Exposition Internationale,* 9:11.

49. Labbé, *Paris, Exposition Internationale,* 9:13.

50. McCloskey, *Artists of World War II,* 94.

51. Laszlo Moholy-Nagy, "The 1937 International Exhibition, Paris," *Architectural Record* 82 (Oct. 1937): 82.

52. Suganami Sohji, "Japan," *Exposition Paris 1937: Arts, Crafts, Sciences in Modern Life* 14 (July–Aug. 1937): 14–15. Unfortunately, the 1940 Grand International Exposition in Tokyo and Yokohama that Suganami was referring to would be canceled due to the conflict with China.

53. Kawahata Naomichi, "Pari Bankoku Hakurankai: Roteishita sho mondai" [The Paris International Exposition: an exposé of various issues], in *"Teikoku" to bijutsu: 1930 nendai Nihon no taigai bijutsu senro* ["Empire" and art: Japanese art of the 1930s and its strategic expansion abroad], ed. Omuka Toshiharu (Tokyo: Kokusho Hankōkai, 2010), 405–44, esp. 421–34.

54. "Paris Held Likely to Continue Fair," *New York Times,* Oct. 24, 1937, 1.

55. Labbé, *Paris, Exposition Internationale* (Paris, 1939), 9:185, 187. See Labbé's earlier reference to Sinclair Lewis's *Babbitt* above, n12.

56. David Littlejohn, "États-Unis," in Lemoine, *Paris 1937,* 156–57.

57. Littlejohn, "États-Unis," 156; "Paris Held Likely," 3.

3. Fantasies of Consumption at Schaffendes Volk: National Socialism and the Four-Year Plan

1. Adam Tooze, in his seminal study of the Nazi war economy, notes that at this time, Germany was twenty-five to thirty years behind the United States in long-term growth. See Adam Tooze, *The Wages of Destruction: The Making and Breaking of the Nazi Economy* (New York: Viking, 2007), 144.

2. Despite the fact that the German economy became increasingly state-

directed under National Socialism, pre-1933 market structures were left largely intact. Jonathan Weisen, *Creating the Nazi Marketplace: Commerce and Consumption in the Third Reich* (Cambridge: Cambridge University Press, 2011), 11.

3. Stefanie Schäfers cites the official report as listing the total number of visitors at 6,904,907 but believes the number was significantly lower than that. She notes that only 4,344,028 tickets were sold and about 50,000 free tickets were given to the press and other visitors. Another 1.2 million tickets were sold just for the amusement park and garden exhibition sections. Stefanie Schäfers, *Vom Werkbund zum Vierjahresplan: Die Ausstellung 'Schaffendes Volk,' Düsseldorf 1937* (Düsseldorf: Droste, 2001), 308.

4. Ernst Maiwald, "Das zeigt deutlich der Vergleich mit Paris," final report, Oct. 14, 1937, Stadtarchiv Düsseldorf, XVIII 1707.

5. Frederick T. Birchall, "Vast German Fair is Built in Secret to Rival Paris Fete: Nation's Biggest Exposition, Laid Out at Duesseldorf, Will Be Opened by Goering May 8," *New York Times*, Apr. 26, 1937.

6. On the problem of too many exhibitions in the 1930s, see for example "Sind wir mit Ausstellungen überfüttert?" *Deutsche Allgemeine Zeitung*, May 12, 1937; and H. Ruban, "Rationalisierung im Ausstellungswesen," *Die Deutsche Volkswirtschaft*, May 1, 1937. On the arrangements by the Werberat, see Schäfers, *Vom Werkbund zum Vierjahresplan*, 96.

7. Jeffrey Schnapp has written that Sironi was responsible for the design of the Italian exhibition at Schaffendes Volk, but I have not found any evidence in the Schaffendes Volk archives or in the archives of the Auswärtiges Amt that this Italian participation in the exhibition took place. See Jeffrey Schnapp, "Flash Memories (Sironi on Exhibit)," *South Central Review* 21, no. 1 (2004): 43. Schnapp's text unfortunately does not cite his source. When I contacted him, he was not able to direct me to the original source for this information.

8. Statistics stated in the final report on the exhibition by Dr. Maiwald, Oct. 14, 1937. Stadtarchiv Düsseldorf, XVIII 1707. There were numerous articles in German newspapers about groups of visitors from various foreign countries to Schaffendes Volk. See, for example, the report on the Belgian Royal Society of Engineers and Industrialists visiting the exhibit in the *Deutsche Bergwerkszeitung*, July 27, 1937; or the article in the *Völkischer Beobachter* (July 10, 1937) on the large numbers of visitors to Schaffendes Volk and the "immense echo" of the exhibition abroad. The number of foreign visitors was also noted by Director

Hattrop in his article, "Ist die Reichsausstellung 'Schaffendes Volk' ein Erfolg?" *Ruhr und Rhein Wirstschaftszeitung* 25, June 18, 1937.

9. On the close ties between the Duke of Windsor, Wallis Simpson, and the Nazi regime as well as the German plots to return him to the British throne, to kidnap him, and other related theories of espionage, see Jonathan Petropoulos, *Royals and the Reich: The Princes von Hessen in Nazi Germany* (Oxford: Oxford University Press, 2006), 206–18.

10. Protocol for royal visit, Sept. 4–5, 1937, Ebel 124, Stadtsarchiv Düsseldorf. See also *Reichsausstellung Schaffendes Volk Düsseldorf 1937. Ein Bericht.* Herausgegeben von Dr. E. W. Maiwald. Zusammengestellt und Bearbeitet von Richard W. Geutebruck, vol. 1 (Düsseldorf: 1939). Maiwald also served as assistant commissioner for the German participation at the 1937 Paris Exposition, and his career shows a pattern of unsettling continuities shared by other Nazi statesmen before and after the Third Reich. In addition to serving as an administrator for numerous Nazi exhibitions, he also served as commissioner for the 1929 Barcelona pavilion, built by Mies van der Rohe, as well as the 1953 exhibit "Alle sollen besser leben" in Düsseldorf. See Schäfers, *Vom Werkbund zum Vierjahresplan*, 404.

11. Paul Betts, *The Authority of Everyday Objects* (Berkeley: University of California Press, 2004), 31.

12. On the role of the Werkbund in the early conception of the exhibit Schaffendes Volk, see Schäfers, *Vom Werkbund zum Vierjahresplan*, 61–103; and on the Gleichhaltung of the Werkbund, see Joan Campbell, *The German Werkbund: The Politics of Reform in the Applied Arts* (Princeton, NJ: Princeton University Press, 1978).

13. Paul Betts, "The Nierentisch Nemesis: Organic Design as West German Pop Culture," *German History* 19, no. 2 (2001): 197. See also Joachim Petsch, "Möbeldesign im Dritten Reich und die Erneuerung des Tischler-Gewerbes seit dem ausgehenden 19. Jahrundert," in *Design in Deutschland 1933–45. Ästhetik und Organisation des Deutschen Werkbundes im "Dritten Reich"*, edited by Sabine Weißler (Berlin: Giessen/Werkbund Archiv, 1990), 43.

14. Despina Stratigakos, "What Is a German Home? Interior Domestic Design and National Identity in the Third Reich," paper presented at the annual meeting of the College Art Association, Los Angeles, Feb. 2012.

15. Stratigakos, "What Is a German Home?"

16. *Das Industrieblatt Stuttgart*, June 15, 1937, 26.

17. *Das Industrieblatt Stuttgart*, June 15, 1937, 23.

18. Schäfers, *Vom Werkbund zum Vierjahresplan*, 314.

19. Hugo Herdeg, *Cahiers d'Art* 8–10 (1937). For a discussion of these exhibitions and the aestheticization of German machinery in French publications, see Fiss, *Grand Illusion*, 107–11.

20. E. Heinson, "Aufbau und Inhalt der Reichsausstellung 'Schaffendes Volk,'" Düsseldorf—Schlageterstadt 1937, in *Stahl und Eisen*. *Zs. für das deutsche Eisenhüttenwesen*, 57. 1937, H. 18, S. 468; cited in Hans-Ulrich Thamer, "Geschichte und Propaganda. Kulturhistorische Ausstellungen in der NS-Zeit," *Geschichte und Gesellschaft* 24, no. 3 (1998): 368.

21. E. W. Maiwald, "'Schaffendes Volk'—ausstellungstechnisch gesehen," Oct. 14, 1937. Stadtarchiv Düsseldorf, XVIII 1707.

22. Hans-Ulrich Thamer similarly argues that Jeffrey Herf's concept of "reactionary modernism" illuminates how "an enthusiasm for technology and progressiveness" could be joined with inhumane racist-based ideology through the implementation of modern aesthetic strategies in exhibition design. See Thamer, "Geschichte und Propaganda," 381.

23. Maiwald, "Paris als die "letzte Weltausstellung," final report, Oct. 14, 1937, Stadtarchiv Düsseldorf, XVIII 1707.

24. Fiss, *Grand Illusion*, 70–98.

25. Peter Fritzsche, "Nazi Modern," *Modernism/Modernity* 3, no. 1 (1996): 7.

26. Schaffendes Volk Pressedienst, May 1937. Ebel 124, Stadtarchiv Düsseldorf.

27. Thamer, "Geschichte und Propaganda," 349–81; Winfried Sonia Hildebrand, "Die Ausstellung als Erlebnis—'Gebt mir vier Jahre Zeit!'" in *100 Jahre Deutscher Werkbund 1907–2007*, ed. Winfried Nerdinger (Munich: Prestel, 2007), 209.

28. Tooze, *Wages of Destruction*, 225.

29. Maiwald, ""Die politische und propagandistische Bedeutung der Ausstellung," final report, Oct. 14, 1937, Stadtarchiv Düsseldorf, XVIII 1707.

30. "Eine Lesitungschau deutscher Arbeit. Zur Eröffnung der Reichsausstellung 'Schaffendes Volk,' Düsseldorf, 1937," *Zeitschrift des Vereines Deutscher Ingenieure*, May 8, 1937.

31. "Göring: Vierjahresplan—Beginn neuen technischen Zeitalters. Die

Rede des Ministerpräsidenten zur Düsseldorfer Ausstellung," *Berliner Tageblatt,* May 9, 1937.

32. The organization of Schaffendes Volk coincided with the establishment of a Reichsforschungsrat, which was intended by the Nazi regime to centralize scientific research. The history of this initiative in relation to Nazi polycratic governance and the Four-Year Plan is too complex to outline in detail here, but as Sören Flachowsky's study demonstrates, engineers and scientists went to great lengths to coordinate their efforts and apply their diverse knowledge fields to the goals of economic self-sufficiency, total war, and genocide. Sören Flachowsky, *Von der Notgemeinschaft zum Reichsforschungsrat: Wissenschaftspolitik im Kontext von Autarkie, Aufrüstung, und Krieg* (Stuttgart: Franz Steiner, 2008).

33. "Sinfonie der Arbeit: Schaffendes Volk—eine Weltereignis," *Berliner Borsen-Zeitung,* May 8, 1937. BA NS5 VI 9926.

34. See Fiss, *Grand Illusion,* 84–98.

35. Tooze, *Wages of Destruction,* 228–29.

36. Schaffendes Volk Pressedienst, "Neue deutsche Werkstoffwirtschaft, Sonderdienst," May 10, 1937. Ebel 124, Stadtarchiv Düsseldorf.

37. "Was Deutschland rechtmäßig gehört! Admiral Rümann eröffnet die Kolonialausstellung in Düsseldorf," *Berliner Börsen-Zeitung,* May 15, 1937.

38. *Schaffendes Volk: die Große Deutsche Ausstellung Düsseldorf-Schlagerterstadt 1937, Mai-Oktober 1937,* dir. Horst Ebel, text by Otto Ernst Wülfing, Stadtarchiv Düsseldorf, 4-59-0 Arnold Emundts.

39. *Schaffendes Volk: die Große Deutsche Ausstellung Düsseldorf-Schlagerterstadt 1937.* Stadtarchiv Düsseldorf, 4–59–0 Arnold Emundts.

40. Professor Dr. Grund (Direktor der Staatlichen Kunstakademie), "Die Schlageterstadt: Zur künstlerischen Gestaltung der Düsseldorfer Ausstellung," *Ruhr und Rhein, Wirtschaftszeitung* 25 (June 18, 1937). BA NS5 VI 9927.

41. Schäfers, *Vom Werkbund zum Vierjahresplan,* 268.

42. "In Mittelpunkt steht der Mensch: Die Frau auf der Reichsausstellung 'Schaffendes Volk' in Düsseldorf—Zahllose Anregungen für die deutsche Hausfrau." *Nationalsozialistische Partei-Korrespondenz* 116 (May 24, 1937).

43. Nancy R. Reagin, *Sweeping the German Nation: Domesticity and National Identity in Germany, 1870–1945* (New York: Cambridge University Press, 2007), 150.

44. Reagin, *Sweeping the German Nation,* 151.

45. Reagin, *Sweeping the German Nation*, 163.

46. Irene Günther, *Nazi Chic?: Fashioning Women in the Third Reich* (Oxford: Berg, 2004), 310, 372–76.

47. Jeff Schutts, "'Die erfrischende Pause': Marketing Coca-Cola in Hitler's Germany," in *Selling Modernity: Advertising in Twentieth-Century Germany*, ed. Pamela E. Swett, S. Jonathan Wiesen, and Jonathan R Zatlin (Durham, NC: Duke University Press, 2007), 170.

48. Schutts, "Die erfrischende Pause," 177–80.

49. Pamela E. Swett, S. Jonathan Wiesen, and Jonathan R. Zatlin, introduction to *Selling Modernity: Advertising in Twentieth-Century Germany* (Durham, NC: Duke University Press, 2007), 14.

50. Corey Ross, "Visions of Prosperity: The Americanization of Advertising in Interwar Germany," in Swett, Wiesen, and Zatlin, *Selling Modernity*, 18.

51. Holm Friebe, "Branding Germany: Hans Domizlaff's Markentechnik and Its Ideological Impact," in Swett, Wiesen, and Zatlin, *Selling Modernity,* 85.

52. Here Friebe is citing Lutz Hachmeister, "Die Welt des Joseph Goebbels," in *Das Goebbels-Experiment: Propaganda und Politik*, ed. Lutz Hachmeister and Michael Kloft (Munich: Deutsche Verlags-Anstalt, 2005), 7.

53. Michael Geyer, "Germany, or, the Twentieth Century as History," *South Atlantic Quarterly* 96, no. 4 (1997): 693. Cited in Alon Confino and Rudy Koshar, "Regimes of Consumer Culture: New Narratives in Twentieth-Century German History," *German History* 19, no. 2 (2001): 159–60.

54. Hartmut Berghoff, "Enticement and Deprivation: The Regulation of Consumption in Pre-War Nazi Germany," in *The Politics of Consumption: Material Culture and Citizenship in Europe and America*, ed. Martin Daunton and Matthew Hilton (Oxford: Berg, 2001), 175.

4. Whose Modernity? Utopia and Commerce at the 1939 New York World's Fair

Epigraph: President William McKinley, speech, Sept. 5, 1901, quoted in the *New York Times*, Sept. 6, 1901, 1. McKinley was assassinated at the Temple of Music at the Pan-American Exposition on September 6.

1. Popular historian Frederick Lewis Allen's *Only Yesterday* (New York: Harper, 1931) describes the prevailing view of the 1920s: "The prestige of science was colossal. The man in the street and the woman in the kitchen . . . were

ready to believe that science could accomplish almost anything" (164–65). This view had remarkable staying power.

2. Arthur Ekirch Jr., *Ideologies and Utopias: The Impact of the New Deal on American Thought* (Chicago: Quadrangle, 1969), 79.

3. Ekirch, *Ideologies and Utopias*, 82.

4. Richard Hofstadter, *The Age of Reform: From Bryan to F.D.R.* (New York: Knopf, 1955), 324–25.

5. Arthur M. Schlesinger Jr., *The Age of Roosevelt*, vol. 2, *The Coming of the New Deal* (Boston: Houghton Mifflin, 1960), 386.

6. Quoted in Robert Kargon and Arthur Molella, *Invented Edens: Techno-Cities of the Twentieth Century* (Cambridge: MIT Press, 2008), 37.

7. Schlesinger, *Age of Roosevelt*, 2:392.

8. Henry Adams, *The Education of Henry Adams* (Boston: Houghton Mifflin, 1918), 343–45.

9. Carol Hagan, "Visions of the City at the 1939 New York World's Fair" (PhD diss., University of Pennyslvania, 2000), 8–18; Joseph Cusker, "The World of Tomorrow: The 1939 New York World's Fair" (PhD diss., Rutgers, 1990), 12–15.

10. Grover Whalen, *Mr. New York: The Autobiography of Grover A. Whalen* (New York: Putnam, 1955), 174–76.

11. Letter, Franklin D. Roosevelt to Franklyn Paris, 9/18/35, F. D. Roosevelt Papers, Roosevelt Library, Hyde Park NY (FDRP), PPF 3000.

12. Frank Morton Todd, *The Story of the Exposition: Being the Official History of the International Celebration Held at San Francisco in 1915* (New York: Putnam, 1921), 109.

13. Memorandum, 16 August 1937, Roosevelt Papers, POF 2147.

14. Lewis Mumford, address, December 11, 1935, New York Public Library MS 2233, Box 918, f.6 "Progressives in the Arts."

15. Manifesto, enclosed in a letter Michael Hare to Robert Moses, 4/27/36 in NYPL MS 2233, Box 136, f.2.

16. George McAneny, speech to Merchants Club, March 4, 1936, NYPL MS 2233, Box 918, f.3; Letter, McAneny to R. Patterson, April 20, 1936, Box 136, f.2; Letter F. Jewett to Wiley Corbett, March 3, 1936, Box 136, f.6.

17. E. Bernays, Speech to the Merchant Association (n.d.), NYPL MS 2233, Box 917, f.8; B. Lichtenberg, "Business Backs New York World Fair to Meet the New Deal Propaganda," *Public Opinion Quarterly* 2, no. 2 (Apr. 1938): 314.

18. John Cawelti, "America on Display: The World's Fairs of 1876, 1893, 1933," in *The Age of Industrialism in America: Essays in Social Structure and Cultural Values*, ed. Frederic Cople Jaher (New York: Free Press, 1968), 348.

19. Roland Marchand, "Corporate Imagery and Popular Education: World's Fairs and Expositions in the United States, 1893–1940," in *Consumption and American Culture*, ed. David Nye and Carl Pedersen (Amsterdam: VU University Press, 1991), 21–25.

20. Henry Elsner Jr., *The Technocrats* (Syracuse, NY: Syracuse University Press, 1967), 23–27.

21. R. D. Kohn, radio talk, Dec. 17, 1936, New York Public Library MS 2233, Box 918, f.3.

22. Robert Kohn, "Social Ideals in a World's Fair," *North American Review* 247, no. 1 (1939): 117. See also Robert Rydell, "The Fan Dance of Science: American World's Fairs in the Great Depression," *Isis* (1985): 525–42; and Peter Kuznick, "Losing the World of Tomorrow: The Battle over the Presentation of Science at the 1939 New York World's Fair," *American Quarterly* 46 (1994): 341–73. Rydell and Kuznick describe some scientists' push for a separate science pavilion. For more favorable contemporary reception see J. M., "Science and the New York World's Fair," *Scientific Monthly* 48 (May 1939): 471–75; and R. L. Duffus, "The Beginning of a World, Not the End," *New York Times Magazine*, July 2, 1939, 1–2.

23. Hand-drawn diagram, NYPL MS 2233 Box 131, Design Progress Report.

24. Grover Whalen, testimony, *Hearings before the Committee on Foreign Affairs, House of Representatives, 75th Congress, on H. J. Res. 234, March 23, 1937* (Washington, DC: Government Printing Office, 1937), 12–13.

25. Letter, FDR to Grover Whalen [unsent, 1937], F. D. Roosevelt Presidential Library, POF 2147. At the urging of his aides this letter was not sent. Mrs. Roosevelt apparently forcefully disagreed. FDR wrote: "Will either one or both of you [Stephen Early, Marvin McIntyre] send a line to the Missus to explain why you have prevented me from sending a letter to Grover Whalen?" *Memo to STE, Mac, 16* August 1937, POF 2147. Mrs. Roosevelt later appeared at the fair wearing a dress printed with a Trylon and Perisphere pattern.

26. M. Woll, testimony, *Hearings before the Committee on Foreign Affairs*, 38–41.

27. Pieter van Wesemael, *Architecture of Instruction and Delight* (Rotterdam: Uitgeverij 010, 2001), 483–89.

28. NYPL MS 2233, Box 134, f.14. "The City of Tomorrow Morning" was Kohn's preferred appellation for *Democracity*, but virtually no one else liked it, and it was shortened to "The City of Tomorrow": "On Mr. Whalen's authority, the word 'Morning' must be dropped from the phrase 'City of Tomorrow Morning.'" Letter, Mrs. Simpson to Mr. Milward, Oct. 13, 1938, NYPL, MS 2233, Box 918, f.3.

29. Wesemael, *Architecture*, 516–17.

30. "Crowds Awed by Fair's Vastness and Medley of Sound and Color," *New York Times*, May 1, 1939, 2.

31. Nicholas Cull, "Overture to an Alliance: British Propaganda at the New York World's Fair 1939–1940," *Journal of British Studies* 36, no. 3 (1997): 334–36.

32. Cull, "Overture," 342, 351.

33. "Japan's Good-Will Broadcast to Fair," *New York Times,* Apr. 3, 1939, 3.

34. William Bernbach and Herman Jaffe, eds., *Book of Nations: New York World's Fair* (New York: Winkler and Kelmans, 1939), 102.

35. Giuseppe Cantu, "Italy's Hope in Fair Pavilion Was to Show World New Aims," *New York Times*, Aug. 28, 1939, 10.

36. James Mauro, *Twilight at the World of Tomorrow: Genius, Madness, Murder, and the 1939 World's Fair on the Brink of War* (New York: Ballantine Books, 2010), 217; Bernbach and Jaffe, *Book of Nations*, 99–100.

37. James Dugan, "Peace Powers at the Fair," *New Masses*, June 27, 1939, 27; Dugan, "Tomorrow's World," *New Masses*, May 2, 1939, 17.

38. Anthony Swift, "The Soviet World of Tomorrow at the New York World's Fair, 1939," *Russian Review* 57, no. 3 (1998): 367–69; Bernbach and Jaffe, *Book of Nations*, 168–69.

39. R. D. Kohn, "Democracity," NYPL MS 2233, Box 137, f.7.

40. *Your World of Tomorrow* (New York: The Fair, 1939), 2.

41. Hagan, "Visions of the City," 73–75.

42. An interesting recollection of *Democracity* can be found in Jeffrey Hart, "The Last Great Fair," *New Criterion* 23 (Jan. 2005): 74–78.

43. *Democracity* script, NYPL MS2233, Box 918, f.3.

44. L. Mumford, "The Sky Line in Flushing," July 29, 1939, reprinted in Lewis Mumford, *Sidewalk Critic*, ed. Robert Wojtowicz (New York: Princeton Architectural Press, 1998), 243. See also Hagan, "Visions of the City," 76–85.

45. F. V. O'Connor, "The Usable Future: The Role of Fantasy in the Promotion of a Consumer Society for Art," in *Dawn of a New Day: The New York*

World's Fair 1939/40, ed. Helen Harrison and Joseph Cusker (New York: NYU Press, 1980), 62.

46. "Fair Called Model for City Planners," *New York Times*, June 25, 1939, 35.

47. Hagan, "Visions of the City," 102–105; Mumford, *Sidewalk Critic*, 246.

48. W. Kaempffert, "Seeing the World of Tomorrow from a Chair Train," *New York Times*, Sept. 10, 1939, D8.

49. General Motors, *Your Guide to General Motors Highways and Horizons Exhibit New York World's Fair 1939* (New York: General Motors, 1939), 2.

50. Roland Marchand, "The Designers Go to the Fair II: Norman Bel Geddes, The General Motors 'Futurama' and the Visit to the Factory Transformed," *Design Issues* 8, no. 2 (1992): 25. See also Marchand, *Creating the Corporate Soul: The Rise of Public Relations and Corporate Imagery in American Big Business* (Berkeley: University of California Press, 1998), 249–311.

51. Marchand, "Designers," 34; Folke Kihlstedt, "Utopia Realized: The World's Fairs of the 1930s," in *Imagining Tomorrow*, ed. Joseph Corn (Cambridge: MIT Press, 1986), 107–108; and Joseph Corn and Brian Horrigan, *Yesterday's Tomorrows: Past Visions of the American Future* (Baltimore, MD: Johns Hopkins University Press, 1996), 46–50.

52. Robert Rydell, *World of Fairs: The Century-of-Progress Expositions* (Chicago: University of Chicago Press, 1993), 135.

53. "America in 1960," *Life*, June 5, 1939, 81, 84.

54. E. B. White, "The World of Tomorrow," in *Essays of E. B. White* (New York: Harper and Row, 1977), 114.

55. Walter Lippmann, "A Day at the World's Fair," *Current History* 50 (July 1939): 50–51.

56. Letter, Kohn to Fordyce, July 6, 1937 and "The City," NYPL MS2233, Box 918, f.7.

57. Kargon and Molella, *Invented Edens*, 24.

58. Mumford, *Sidewalk Critic*, 242–43.

59. Archer Winsten, "The City Goes to the Fair," *New York Post*, June 23, 1939.

60. This film is available for viewing (as of January 2015) at http://archive .org/details/ToNewHor1940.

61. Sinclair Lewis, *Babbitt* (New York: New American Library, 1961 [1922]), 8.

62. This film is available for viewing (as of January 2015) at http://archive.org/details/middleton_family_worlds_fair_1939. See also Ethan Robey, "The Battle of the Centuries," http://exhibitions.nypl.org/biblion/worldsfair/enter-world-tomorrow-futurama-and-beyond/essay/essay-robey-dishwashing; William Bird, "Enterprise and Meaning: Sponsored Film, 1939–1949," *History Today*, Dec. 1989, 28–29.

63. Marco Duranti, "Utopia, Nostalgia and World War at the 1939–40 New York World's Fair," *Journal of Contemporary History* 41, no. 4 (2006): 674.

64. Francis Edmonds Tyng, *Making a World's Fair* (New York: Vantage, 1958), 101.

65. Duranti, "Utopia, Nostalgia and World War," 682.

66. Sheldon Reaven, "New Frontiers: Science and Technology at the Fair," 96, and Rosemarie H. Bletter, "'The Laissez-Fair,' Good Taste and Money Trees," 119–22, both in *Remembering the Future: The New York World's Fair from 1939 to 1964*, ed. Rosemarie Haag Bletter (New York: Rizzoli, 1989); General Motors Promotional Video, *Futurama II*, accessed Jan. 2015, http://www.youtube.com/watch?v=2–5aK0H05jk.

67. Adams, *Education*, 343–45.

5. Modernity on Display: The 1940 Grand International Exposition of Japan

1. Jeffrey Herf, *Reactionary Modernism: Technology, Culture, and Politics in Weimar and the Third Reich* (Cambridge: Cambridge University Press, 1984).

2. Association of the Japan International Exposition, *Grand International Exposition of Japan, March–August 1940: In Celebration of the 2600th Anniversary of the Accession of the First Emperor Jimmu to the Throne, the Purport, Aims and Outline of the Exposition* (Tokyo: Association of the Japan International Exposition, 1937), 5.

3. Hiromi Mizuno, *Science for the Empire: Scientific Nationalism in Modern Japan* (Stanford, CA: Stanford University Press, 2009), 144.

4. Mizuno, *Science for the Empire*, 144.

5. Association of the Japan International Exposition, *Grand International Exposition of Japan*, 11–12.

6. Association of the Japan International Exposition, *Grand International Exposition of Japan*, 3–4.

7. Association of the Japan International Exposition, *Grand International Exposition of Japan*, xi–xii.

8. Ginjirō Fujihara, *The Spirit of Japanese Industry* (Tokyo: Hokuseido Press, 1936), vi.

9. Fujihara, *Spirit of Japanese Industry*, x–xi.

10. Commemorative Association for the Japan World Exposition, *Official Report of the Japan World Exposition, Osaka, 1970*, vol. 1 (Suita City, Osaka: Commemorative Association for the Japan World Exposition, 1970), 31.

11. "International Exposition of Japan in 1940 to be held at Tokyo and Yokohama to Commemorate 2,600th Year of Founding of Japanese Empire," *Japan Trade Review* 11, no. 2 (March 1938): 37–44, esp. 37.

12. Association of the Japan International Exposition, *Grand International Exposition*, 7.

13. Association of the Japan International Exposition, *Grand International Exposition*, 7.

14. Association of the Japan International Exposition, *Grand International Exposition*, 10.

15. Commemorative Association for the Japan World Exposition, *Official Report*, 31.

16. Association of the Japan International Exposition, *Grand International Exposition*, 14.

17. Commemorative Association for the Japan World Exposition, *Official Report*, 31.

18. Association of the Japan International Exposition, *Grand International Exposition*, 13–14.

19. Association of the Japan International Exposition, *Grand International Exposition*, 15.

20. Jonathan M. Reynolds, *Maekawa Kunio and the Emergence of Japanese Modernist Architecture* (Berkeley: University of California Press, 2001), 121.

21. Yoshimi Shunya, *Hakurankai no seijigaku* [The politics of expositions] (Tokyo: Chūō Kōronsha, 1992), 216.

22. Association of the Japan International Exposition, *Grand International Exposition*, 15.

23. Association of the Japan International Exposition, *Grand International Exposition*, 16.

24. Association of the Japan International Exposition, *Grand International Exposition*, 16.

25. Barak Kushner, *The Thought War: Japanese Imperial Propaganda* (Honolulu: University of Hawaii Press, 2006), 39.

26. Edward Seidensticker, *Tokyo Rising: The City since the Great Earthquake* (New York: Alfred A. Knopf, 1990), 102–103.

27. Commemorative Association for the Japan World Exposition, *Official Report*, 31.

28. "Shinsō nareru banpaku jimukyoku o rikugun shōbyōhei shūyōjo ni katsuyō" [Refurbished expo headquarters put to practical use as a care facility for sick and wounded soldiers], *Banpaku* 30 (Nov. 1938): 4–5.

29. Daqing Yang, *Technology of Empire: Telecommunications and Japanese Expansion in Asia, 1883–1945* (Cambridge, MA: Harvard University Asia Center, 2010), 158.

30. Yang, *Technology of Empire*, 204.

31. Mizukoshi Shin, "Social Imagination and Industrial Formation of Television in Japan," *Bulletin of the Institute of Socio-Information and Communication Studies, University of Tokyo* (March 1996): 1–14.

32. Association of the Japan International Exposition, *Grand International Exposition*, 16.

33. Shinanyaku, *Moboroshi no 1940 nen keikaku: Taiheiyō sensō no zenya, 'kiseki no toshi' ga tanjō shita* [The dream plan of 1940: the birth of the "City of Wonder" just before the Pacific War] (Tokyo: Asupekuto, 2009), 109.

34. Association of the Japan International Exposition, *Grand International Exposition*, 41.

35. "A Historical Outline of the Development of Manchurian Railways," *Contemporary Manchuria* 3, no. 3 (July 1939): 36–61. Also see "Manchuria's Super-Express 'Asia,'" *Contemporary Manchuria* 2, no. 1 (1938): 45–60.

36. Nicholas Mirzoeff, *An Introduction to Visual Culture* (London: Routledge, 1999), 29.

37. Yang, *Technology of Empire*, 158.

38. Mizuno, *Science for the Empire*, 4.

39. Yasuko Suga, "Modernism, Nationalism and Gender: Crafting 'Modern' Japonisme," *Journal of Design History* 21, no. 3 (2008): 259–75, esp. 259.

40. Iwao Yamawaki, "Reminiscences of Dessau," *Design Issues* 2, no. 2 (1985): 56–68, esp. 56–57.

41. Akiko Takenaka, "The Construction of a Wartime National Identity:

The Japanese Pavilion at New York World's Fair 1939/40" (master's thesis, MIT, 1997), 31.

42. Bill Cotter, *The 1939–1940 New York World's Fair* (Charleston, SC: Arcadia Publishing, 2009), 59.

43. Yoshida Mitsukuni, *Bankoku hakurankai: Gijutsu benmei shiteki ni* [International expositions: from a history of technological civilization perspective] (Tokyo: Nihon Hōsō Kyōkai, 1985), 204.

44. Takenaka, "Construction," 31.

45. Andrew S. Dolkart, "The Skyscraper City," *The Architecture and Development of New York City* (2004), 12, accessed June 14, 2012, http://ci.columbia.edu/0240s/0242_3/0242_3_fulltext.pdf.

46. Gennifer Weisenfeld, "Touring Japan-as-Museum: *NIPPON* and Other Japanese Imperialist Travelogues," *Positions* 8, no. 3 (2000): 747–93, esp. 748.

47. Andrea Germer, "Visual Propaganda in Wartime East Asia: The Case of Natori Yōnosuke," *Asia-Pacific Journal* 9, no. 3 (2011), accessed June 17, 2012, http://www.japanfocus.org/-Andrea-Germer/3530.

48. Gennifer Weisenfeld, "Publicity and Propaganda in 1930s Japan: Modernism as Method," *Design Issues* 25, no. 4 (2009): 13–28, esp. 22.

49. Weisenfeld, "Publicity and Propaganda," 23.

50. Andrea Germer, "Artists and Wartime Politics: Natori Yōnosuke—A Japanese Riefenstahl?," *Journal of the German Institute for Japanese Studies* 24 (2012): 21–50; Germer, "Visual Propaganda."

51. Gennifer Weisenfeld, "The Expanding Arts of the Interwar Period," in *Since Meiji: Perspectives on the Japanese Visual Arts, 1868–2000*, ed. J. Thomas Rimer, with translations by Toshiko McCallum (Honolulu: University of Hawaii Press, 2012), 66–98, esp. 94–95.

52. Atsushi Shibasaki, "Activities and Discourses on International Cultural Relations in Modern Japan: The Making of KBS (Kokusai Bunka Shinko Kai), 1934–53," *Journal of Global Media Studies* 8 (Mar. 2011): 25–41.

53. Germer, "Artists and Wartime Politics," 30.

54. "Cultural Accord Aims at Exchange of Films, Books," *Japan Times and Mail*, Nov. 25, 1938, 1.

55. Taikan Yokoyama, "Western and Oriental Art Is Discussed by Yokoyama," *Japan Times and Mail*, Nov. 25, 1938, 2.

56. Naofumi Masumoto, "Interpretations of the Filmed Body: An Analysis of the Japanese Version of Leni Riefenstahl's *Olympia*," in *Critical Reflections on*

Olympic Ideology: Second International Symposium for Olympic Research, ed. Robert K. Barney and Klaus V. Meier (London, Ontario: Centre for Olympic Studies, University of Western Ontario, 1994), 146–57.

57. Yang, *Technology of Empire*, 158.

58. Robert S. Schwantes, "Japan's Cultural Foreign Policies," in *Japan's Foreign Policy, 1868–1941: A Research Guide*, ed. James W. Morley (New York: Columbia University Press, 1974), 153–83, esp. 177–78.

59. "300 Japanese Folk Art Works to Be Exhibited in New York," *Japan Times and Mail*, Nov. 29, 1938, 1.

60. Hiroshi Watanabe, *The Architecture of Tokyo: An Architectural History in 571 Individual Presentations* (Stuttgart: Edition Axel Menges, 2001), 80; Arata Isozaki, *Japan-ness in Architecture*, trans. Sabu Kohso, ed. David B. Stewart (Cambridge: MIT Press, 2011), 14.

61. Bruno Taut, *Fundamentals of Japanese Architecture* (Tokyo: Kokusai Bunka Shinkokai, 1936), 9.

62. Sandra Kaji-O'Grady, "Authentic Japanese Architecture after Bruno Taut: The Problem of Eclecticism," *Fabrications* 11, no. 2 (2001): 1–12, esp. 2.

63. Tetsuo Najita and H. D. Harootunian, "Japanese Revolt against the West: Political and Cultural Criticism in the Twentieth Century," in *The Cambridge History of Japan*, vol. 6, *The Twentieth Century*, ed. Peter Duus (Cambridge: Cambridge University Press), 711–74, esp. 712.

64. Peter McNeil, "Myths of Modernism: Japanese Architecture, Interior Design and the West, c. 1920–1940," *Journal of Design History* 5, no. 4 (1992): 281–94, esp. 281.

65. Preface to *Antonin Raymond, His Works in Japan, 1920–35* (1938), cited in McNeil, "Myths of Modernism," 293.

66. Burritt Sabin, "Echoes of the Footfalls of Soldiers," *Japan Times*, Dec. 12, 2008, accessed Jan. 21, 2013, http://www.japantimes.co.jp/text/fv20081212a1.html.

67. "Honkan (Japanese Gallery) Floor Map," *Tokyo National Museum*, accessed Jan. 21, 2013, http://www.tnm.jp/modules/r_free_page/index.php?id=115&lang=en.

68. Sutemi Horiguchi, trans. Robin Thompson, "'Japanese Taste' in Modern Architecture," *Art in Translation* 4, no. 4 (2012): 407–34, esp. 410. Originally published as Horiguchi Sutemi, "Gendai kenchiku ni arawareta Nihon shumi ni tsuite," *Shisō* (Jan. 1932).

69. Horiguchi, "Japanese Taste," 410.

70. David B. Stewart, *The Making of a Modern Japanese Architecture: From the Founders to Shinohara and Isozaki* (Tokyo: Kodansha International, 2002), 172.

71. Jonathan M. Reynolds, "Ise Shrine and a Modernist Construction of Japanese Tradition," *Art Bulletin* 83, no. 2 (June 2001): 316–41, esp. 323.

72. Reynolds, *Maekawa Kunio*, 126.

73. Tange Kenzō, "Dai tōa kyōeiken ni okeru kaiin no yōbō" [Member's plea in regard to the Greater East Asia Co-prosperity Sphere], *Kenchiku zasshi* [Architectural journal] 56, no. 690 (Sept. 1942): 744. Cited in Hyunjung Cho, "Hiroshima Peace Memorial Park and the Making of Japanese Postwar Architecture," *Journal of Architectural Education* 66, no. 1 (2012): 72–83, esp. 74.

74. Reynolds, *Maekawa Kunio*, 129.

75. Kevin Michael Doak, *Dreams of Difference: The Japan Romantic School and the Crisis of Modernity* (Berkeley: University of California Press, 1994), 134–35.

76. Paul Valéry, "The European Spirit" (1935), in *History and Politics*, vol. 10 of *The Collected Works of Paul Valéry*, trans. Denise Folliot and Jackson Mathews (London: Routledge and Kegan Paul, 1963), 326–28.

77. H. L. Wesseling, "From Cultural Historian to Cultural Critic: Johan Huizinga and the Spirit of the 1930s," *European Review* 10, no. 4 (Oct. 2002): 485–99, esp. 491.

78. Paul Valéry, "L'Esprit européen," *Les Nouvelles Littéraires*, Nov. 16, 1935, n.p., cited in Valéry, *History and Politics*, 597.

79. Valéry, *History and Politics*, 327.

80. Johan Huizinga, *In the Shadow of To-Morrow: A Diagnosis of the Spiritual Distemper of Our Time* (London: William Heinemann, 1936), 51.

81. Mizuno, *Science for the Empire*, 2.

82. Richard F. Calichman, ed. and trans., *Overcoming Modernity: Cultural Identity in Wartime Japan* (New York: Columbia University Press, 2008). Calichman's translation needs to be read in conjunction with his earlier translation of the writings of Takeuchi Yoshimi, which were published in 2005. See Calichman, ed. and trans., *What Is Modernity?: Writings of Takeuchi Yoshimi* (New York: Columbia University Press, 2005).

83. Calichman, *Overcoming Modernity*, 94.

84. Calichman, *Overcoming Modernity*, 111.

85. Calichman, *Overcoming Modernity*, 123.

86. Calichman, *Overcoming Modernity*, 149.

87. Calichman, *Overcoming Modernity*, 179.

88. Calichman, *Overcoming Modernity*, 180.

89. Calichman, *Overcoming Modernity*, 199.

90. Najita and Harootunian, "Japanese Revolt," 765, 767.

91. Najita and Harootunian, "Japanese Revolt."

92. Michael A. Barnhart, *Japan Prepares for Total War: The Search for Economic Security, 1919–1941* (Ithaca, NY: Cornell University Press, 1987).

93. Marilyn Ivy, "Foreword: Fascism, Yet?" in *The Culture of Japanese Fascism*, ed. Alan Tansman (Durham, NC: Duke University Press, 2009), vii–xii, esp. x.

94. Alan Tansman, "Introduction: The Culture of Japanese Fascism," in Tansman, *Culture of Japanese Fascism*, 1–28.

95. Jonathan M. Reynolds, "Teaching Architectural History in Japan: Building a Context for Contemporary Practice," *Journal of the Society of Architectural Historians* 61, no. 4 (Dec. 2002): 530–36, esp. 532.

96. Reynolds, "Ise Shrine," 325.

97. Reynolds, "Ise Shrine," 316.

98. Yumiko Iida, *Rethinking Identity in Modern Japan: Nationalism as Aesthetics* (London: Routledge, 2002), 60.

99. Iida, *Rethinking Identity*, 66.

6. EUR: Mussolini's Appian Way to Modernity

1. "EUR" and "E42" are sometimes used interchangeably. To avoid confusion, E42 will be used for the fair itself and EUR will be reserved for the fair's district.

2. Vincent Scully, "Louis I. Kahn and the Ruins of Rome," *Museum of Modern Art Members Quarterly*, summer 1992, 3; Archivio Centrale dello Stato, Esposizione Universale di Roma, E42, hereafter E42 Archives. An indispensable guide to the fair and to the E42 collections at the Italian state archives is Tullio Gregory and Achille Tartaro, eds., *E42: Utopia e Scenario del Regime*, 2 vols. (Venice: Cataloghi Marsilio, 1987), the companion catalog for an exhibition marking the fiftieth anniversary of planning for E42. I am indebted to Dr. Carlotta Darò, École nationale supérieure d'architecture Paris Malaquais, and to Dr. Mariapina di Simone, Archivio centrale dello stato, for their assistance in navigating the archives.

3. Giorgio Ciucci, "The Classicism of the E42: Between Modernity and Tradition," trans. Jessica Levine, *Assemblage* 8 (Feb. 1989): 80.

4. Anna Notaro, "Exhibiting the New Mussolinian City: Memories of Empire in the World Exhibition of Rome (EUR)," *Geojournal* 51 (2001): 19. On the Fascist paradox of blending past and future, see Mark Antliff, "Fascism, Modernism, and Modernity," *Art Bulletin* 84 (Mar. 2002): 148–69.

5. Joshua Arthurs, *Excavating Modernity: The Roman Past in Fascist Italy* (Ithaca, NY: Cornell University Press, 2012); Jan Nelis, "Constructing Fascist Identity: Benito Mussolini and the Myth of 'Romanità,'" *Classical World* 100, no. 4 (2007): 391–415.

6. Roger Griffin, *The Nature of Fascism* (New York: St. Martin's, 1991). More recently, Griffin writes, "Fascism is . . . an ideology deeply bound up with modernization and modernity . . . [that] has drawn on a wide range of intellectual currents, both left and right, anti-modern and pro-modern." See "The Palingenetic Core of Fascist Ideology," in *Che cos'è il fascismo? Interpretazioni e prospective di recherche*, ed. A. Campi (Rome: Ideazione editrice, 2003), 97–122, http://www.libraryofsocialscience.com/ideologies/docs/the-palingenetic-core -of-generic-fascist-ideology/index.html.

7. From some perspectives the two structures looked virtually identical, which led to the charges of plagiarism. Their shapes, however, were fundamentally different: the Arco was of a semi-circular design, whereas Saarinen's took the innovative shape of a catenary, of which the Finnish-born master was particularly proud. Personal communication with Mina Marefat, Georgetown University.

8. While never built, in the years running up to the fair, the Arco dell'impero was heavily pre-advertised in film and the printed media, including posters, newspapers, and glossy magazines, for example, *L'Illustrazione Italiana* (special issue, 1942).

9. Joshua Arthurs argues that "Fascist culture" should be taken seriously and on its own terms rather than dismissed as empty propaganda. *Excavating Modernity*, 3–4, and passim. In placing E42's Fascist modernity within the general intellectual and cultural discourse of the era, this chapter builds on Arthurs's thesis.

10. How Fascists reconciled the antihistorical cult of "action" with the promotion of the study of the past is explained by Arthurs, *Excavating Modernity*, 34.

11. For the important role of exhibitions in the Fascist revisioning of Rome,

see Aristotle Kallis, "Fascism in *Mostra*: Exhibitions as Heterotopias," in *The Third Rome, 1922–43: The Making of the Fascist Capital* (New York: Palgrave Macmillan, 2014), 198–225. Also see Marla Stone, "A Fascist Theme Park," in *Visual Sense: A Cultural Reader*, ed. Elizabeth Edwards and Kaushik Bhaumik (New York: Berg, 2008), 271.

12. On exhibitions as Fascist policy, see Claudio Fogu, *The Historic Imaginary: Politics of History in Fascist Italy* (Toronto: University of Toronto Press, 2003), chapter 5; Jeffrey T. Schnapp, "Epic Demonstrations: Fascist Modernity and the 1932 Exhibition of the Fascist Revolution," in *Fascism, Aesthetics, and Culture*, ed. Richard J. Golsan (Hanover, NH: University Press of New England, 1992), 2–37; Diane Ghirardo, "Architects, Exhibitions, and the Politics of Culture in Fascist Italy," *Journal of Architectural Education* 45, no. 2 (Feb. 1992): 67–74.

13. Spiro Kostoff, *The Third Rome, 1870–1950: Traffic and Glory* (Berkeley: University Art Museum, 1973), 37.

14. Notaro, "Exhibiting the New Mussolinian City," 19. We have found no archival evidence that other countries were actually planning their own national pavilions, or were even invited by Italy to do so. The economic sanctions on Italy after its conquest of Addis Abbaba probably discouraged international participation. However, at least a few Americans offered ideas for the Italian pavilion's science displays.

15. Kostoff, *Third Rome*, 37.

16. Ciucci, "Classicism of the E42," 80.

17. Notaro, "Exhibiting the New Mussolinian City," 19.

18. On Il Duce's theatricality, see Marla Stone, "Staging Fascism: The Exhibition of the Fascist Revolution," *Journal of Contemporary History* 28 (Apr. 1993): 215–43.

19. Diane Ghirardo, "Città Fascista: Surveillance and Spectacle," in "The Aesthetics of Fascism," special issue, *Journal of Contemporary History* 31, no. 2 (Apr. 1996): 347, 350, 352.

20. Ghirardo, "Città Fascista," 347–52.

21. Don Peretz, *The Middle East Today*, 6th ed. (Westport, CT: Praeger, 1994), 155, describes Ataturk's policy. For Reza Shah, see Mina Marefat, "Building to Power: Architecture of Tehran, 1921–1941" (PhD diss., MIT, 1988). Marefat notes that Reza Shah "was driven by the same forces as other leaders and similar nationalistic visions" (73). The Shah's program strikingly paralleled Mussolini's.

22. Herf, *Reactionary Modernism*.

23. See Sandro Bellassai, "The Masculine Mystique: Anti-Modernism and Virility in Fascist Italy," *Journal of Modern Italian Studies* 10, no. 3 (2005): 314–35; Francesco Cassata, *Building the New Man: Eugenics, Racial Science and Genetics in Twentieth-Century Italy*, trans. Erin O'Loughlin (Budapest: Central European Press, 2011), which places the New Man in the context of racism and the pseudoscience of eugenics.

24. Victoria de Grazia, *The Culture of Consent: Mass Organization of Leisure in Fascist Italy* (Cambridge, UK: Cambridge University Press, 1981), 201.

25. Quoted in de Grazia, *Culture of Consent*, 191.

26. In its futuristic enthusiasm, it looked like an Italian prototype for Walt Disney's EPCOT, Walt's futuristic vision of a technology-driven America.

27. Despite these unreal expectations, EUR came into being. A somewhat sterile but not unpleasant place, EUR is today a lively business district, with new construction underway, including buildings, parks, and lakes—still "fulfilling" Il Duce's dream of a futuristic green utopia.

28. Ciucci, "Classicism of the E42," 81.

29. Kargon and Molella, *Invented Edens*, chapter 3.

30. Antonioni's film, one of his earliest efforts, was titled *Seven Reeds, One Dress (Sette canne, un vestito)*. Kargon and Molella, *Invented Edens*, 62–65.

31. As illustrated also by Ferrara, discussed in Ghirardo, "Città Fascista," n3.

32. Quoted from Luigi Di Majo and Italo Insolera, *L'EUR e Roma dagli anni Trenta al Duemila* (Rome: Editori Laterza, 1986), 47.

33. For the battle between architects favoring the international style and Fascist modernists in the building of EUR, see Ghirardo, "Architects, Exhibitions," 73–74; and Ciucci, "Classicism of the E42," 83–86.

34. Kostoff, *Third Rome*, 74. We might ask why the Fascist style persisted in EUR after the fall of the regime. One has to assume that in the 1950s classical modernism "survived" rather than "perpetuated" Fascism—that its cultural and stylistic roots went deeper than politics. However, as we shall see below, there was a cultural revolt against it by the 1960s.

35. See Diane Ghirardo, "Italian Architects and Fascist Politics: An Evaluation of the Rationalist's Role in Regime Building," *Journal of the Society of Architectural Historians* 39, no. 2 (1980): 109–27.

36. Such as Gino Severini. Mosaics decorating the central fountain are pictured in Carlo Bertilaccio and Francesco Innamorati, *EUR, E42 Heritage: A Use-*

ful Handbook (Rome: EUR SpA, 2005), 28. We thank Sage Snider for providing a copy of this handbook.

37. Sage Snider, "Realizing the Fascist Vision: Mussolini's Construction of Roman History at EUR" (undergraduate thesis, Yale University, 2012), interprets various monuments and sculptures in EUR.

38. No one using the E42 archives in that imposing colonnaded structure can escape the eerie feeling of going back in time to Mussolini's white city.

39. Kenneth Frampton, *Modern Architecture: A Critical History*, 4th ed. (New York: Thames and Hudson, 2007), 203–204.

40. Mina Marefat, personal communication.

41. For example, F. T. Marinetti, the chief founder of futurism, wrote "Il Poema di Torre Viscosa," a paean to the new Italian city opened by Mussolini in 1938.

42. Scully, "Kahn and the Ruins of Rome," 3.

43. One of first popular magazines to publicize details of the fair was *L'Illustrazione Italiana*, in its December 1938 issue.

44. Nervi et al. to Cini, May 4, 1938, E42 Archives, Box 308, folder 4931/11.

45. E42 Archives, Box 308: 4931.

46. Ghirardo, "Città Fascista," 350. No one really knows what qualified a Roman notable to receive such an accolade. Harry Rand, Smithsonian Institution, personal communication. Clearly, in Mussolini's case, no one dared question his eligibility.

47. E42 Archives, Box 308: 4931/9 and 11.

48. For example, studies submitted by Vincenzo di Berardino, who appears to have worked with Nervi on the drawings, "Arco Monumental E42" E42 Archives, Box 647: 6972. Nervi on soundings, Box 308:4931/10.

49. Letter dated April 28, 1939, E42 Archives, Box 308: 4931/9. The reverse arch was far from unfeasible, but a well-tested technique used in many ancient buildings and forts. Rand, personal communication.

50. Ferretti to Cini, June 26, 1939, E42 Archives, Box 308: 4931/4.

51. While this claim by the proposing engineers is not stated in the clearest Italian, the expression of symbolic value is clear. E42 Archives, Box 308: 4931/11.

52. Voltaggio to Cini, April 19, 1940, E42 Archives, Box 308: 4931.

53. E42 Archives, Box 1003: 9769/38.

54. Borden W. Painter Jr., *Mussolini's Rome: Rebuilding the Eternal City* (New York: Palgrave MacMillan, 2005), xvi.

55. Sealed by the Lateran Treaty of 1929. Courting the support of the Catholic Church for his Fascist regime, the fiercely anti-clerical Mussolini made a tactical peace with the Vatican.

56. Sage Snider suggested this plausible comparison in a personal communication in 2012. See note 36 above.

57. See Schnapp, "Fascism's Museum in Motion," *Journal of Architectural Education* 45, no. 2 (1992): 87–97; Schnapp, "Epic Demonstrations"; Stone, "Staging Fascism"; Ghirardo, "Architects, Exhibitions."

58. Stone, "Staging Fascism," 220–21.

59. Libero Andreotti, "The Techno-Aesthetics of Shock: Mario Sironi and Italian Fascism," *Grey Room* 38 (2010): 38–61.

60. Painter, *Mussolini's Rome*, 78–79.

61. A detailed account of the exhibition's origins, ideology, content, layout, and contribution to *romanità* is provided by Arthurs, *Excavating Modernity*, 94–124.

62. The following description of the museum is based on its present-day configuration. However, with the exception of the addition of the planetarium, there are no signs that the displays have materially changed.

63. Ghirardo notes that such martial rhetoric was typical of Mussolini's propaganda campaigns. Diane Ghirardo, *Building New Communities: New Deal America and Fascist Italy* (Princeton, NJ: Princeton University Press, 1989), 27.

64. The modelmaker Pierino di Carlo began work on it in 1933, only four years after the Lateran accords between Mussolini and the Pope. His creation is a technical and artistic tour de force.

65. Denis Mack Smith, *Mussolini* (New York: Knopf, 1982), 134.

66. The plea was made by Giulio Terzaghi. A copy of the January 1941 issue of the journal with a cover note to Cini was found in E42 Archives, Box 1010:9770/2/35.

67. Paolo Galluzzi, "La storia della scienza nell'E42," in Gregory and Tartaro, *E42: Utopia e Scenario del Regime*, 1:53.

68. "La Mostra di Leonardo e delle Invenzioni" is the lead article of the January 1938 (Fascist year XVI) of the journal *Ingegni e Congegni* [Engineers and devices], subtitled *Le Attualità Scientifiche* [Science now]. It describes itself as the official journal of the National Association of Fascist Inventors. Found in E42 Archives, Box 1010: 9770/2/22.

69. Visco to Cini, August 14, 1939, E42 Archives, Box 1010: 9770/2/22.

70. Galluzzi, "Storia della scienza," 54. Galluzzi also notes that Mussolini took part in the first national exhibition on the history of science in Florence in 1930, which generated much Italian scholarship in the history of science.

71. Enrico Luciana to Visco, March 7, 1939. E42 Archives, Box 1010: 9770/2/23.

72. Galluzzi, "Storia della scienza," 54.

73. Mack Smith, *Mussolini*, 134.

74. E42 Archives, Box 1025: 9770/18. And on invention: proposal of M. M. Apollon, May 7, 1940. Box 1010: 9770/2/32.

75. Mack Smith, *Mussolini*, 134.

76. Box 1047, E42 Archives, is devoted primarily to E42's planned science exhibitions.

77. E42 Archives, Box 794: 7066/1/21.

78. E42 Archives, Box 1047: 9981/1.

79. The architecture and design may seem over the top, even inappropriate for a science museum, but many science and technology museums in Europe and America occupy similarly ornate neoclassical structures.

80. Galluzzi, "Storia della scienza," 61ff.

81. E42 Archives, Box 1025, is devoted to this projected alternative.

82. Gerald Holton, *Science and Anti-Science* (Cambridge, MA: Harvard University Press, 1993), 181.

83. Gregory and Tartaro, *E42: Utopia e Scenario del Regime*; Carlo Carli, Gianni Mercurio, and Luigi Prisco, eds., *E42 EUR: Segno e Sogno del Novecento* (Rome: DataArs, c. 2005).

84. See Painter, *Mussolini's Rome*.

85. As quoted by Mack Smith, *Mussolini*, 136.

86. Mack Smith, *Mussolini*, 137.

87. Ghirardo, *Building New Communities*, 9.

88. Fascist architecture has since acquired a certain retro cachet and attracted growing scholarly interest.

7. Redefining Cultural Conflict: The First Postwar International Exposition

1. Jeffrey Herf, "Reactionary Modernism Reconsidered," in *The Intellectual Revolt against Liberal Democracy 1870–1945*, ed. Z. Sternhell (Jerusalem: Israel Academy, 1996), 133.

2. Brigitte Schroeder-Gudehus and Anne Rasmussen, *Les Fastes du Progrès: Le guide des Expositions universelles 1851–1992* (Paris: Flammarion, 1992), 206.

3. Robert Haddow, "Material Culture and the Cold War: International Trade Fairs and the American Pavilion at the 1958 Brussels World's Fair" (PhD diss., University of Minnesota, 1994), 136. See also Haddow, *Pavilions of Plenty: Exhibiting American Culture Abroad in the 1950s* (Washington, DC: Smithsonian Institution, 1997), chapter 4.

4. K. G. Beauchamp, *Exhibiting Electricity* (London: Institution of Electrical Engineers, 1997), 277–81.

5. Greg Castillo, "Making a Spectacle of Restraint: The Deutschland Pavilion at the 1958 Brussels Exposition," *Journal of Contemporary History* 47 (2012): 118. See also Rika Devos, "Power, Nationalism and National Representation in Modern Architecture and Exhibition Design at Expo 58," in *Nationalism and Architecture*, ed. Raymond Quek and Darren Deane (Farnham, Surrey: Ashgate England, 2012), 81–94.

6. Alain de Botton, *The Architecture of Happiness* (New York: Vintage, 2008), 91–93; Castillo, "Making a Spectacle," 100.

7. See Lewis Siegelbaum, "Sputnik Goes to Brussels: The Exhibition of a Soviet Technological Wonder," *Journal of Contemporary History* 47, no. 1 (2012): 120–36. See also Susan Reid, "The Soviet Pavilion at Brussels '58: Convergence, Conversion, Critical Assimilation or Transculturism?" (Cold War History Project, working paper no. 62, Woodrow Wilson Center for Scholars, Dec. 2010).

8. Walter Waggoner, "U.S. versus U.S.S.R. at Brussels Fair Too," *New York Times*, Apr. 27, 1958, E6.

9. Marquis Childs, *Washington Post*, June 24, 1958, A16.

10. Howard Taubman, "Brussels: American Mistakes and Lessons," *New York Times*, June 1, 1958, SM 11, 14, 16.

11. Taubman, "Brussels," 14.

12. Rydell, *World of Fairs*, 194.

13. Rydell, *World of Fairs*, 14–16; Haddow, *Pavilions*, 98–99.

14. Rydell, *World of Fairs*, 199.

15. Haddow, *Pavilions*, 104–105.

16. Walter Hixson, *Parting the Curtain: Propaganda, Culture and the Cold War 1945–1961* (New York: St. Martin's, 1997), 142–43.

17. *This Is America: Official United States Guide Book Brussels World's Fair 1958*

(N.p.: Office of the Commissioner General, 1958), passim; Haddow, *Pavilions*, 106–10, 174–87; Michael Krenn, "Unfinished Business: Segregation and U.S. Diplomacy at the 1958 World's Fair," *Diplomatic History* 20, no. 4 (1996): 591–612.

18. Relman Morin, "U.S. Exhibit at Brussels Stirs Varied Comment," *Los Angeles Times*, July 13, 1958, A34.

19. Art Buchwald, "U.S. at the Fair," *Los Angeles Times*, May 30, 1958, B5. Cf. Karl Marx's point that "Hegel remarks somewhere that all great world-historic facts and personages appear, so to speak, twice. He forgot to add: the first time as tragedy, the second time as farce." *The Eighteenth Brumaire of Louis Bonaparte*, 1.

20. President's Science Advisory Committee, *Introduction to Outer Space* (Washington, DC: Government Printing Office, 1958), 1. Cf. "To boldly go where no man has gone before," a phrase made popular since 1966 by the television program *Star Trek* and its reruns.

BIBLIOGRAPHY

Archives

Bundesarchiv, Berlin

Franklin D. Roosevelt Papers, Roosevelt Library, Hyde Park NY. PPF 3000, POF 2147

New York World's Fair 1939 and 1940 Incorporated records, New York Public Library MS2233

Stadtarchiv Düsseldorf

Published Works

Adams, Henry. *The Education of Henry Adams*. Boston: Houghton Mifflin, 1918.

Ades, Dawn. "Paris 1937: Art and the Power of Nations." In *Art and Power: Europe under the Dictators 1930–45*, 58–62. London: Thames and Hudson, 1995.

Alfieri, Dino. "Scopo, carattere, significato della Mostra Mussolini e la Rivoluzione." In *Guida della mostra della rivoluzione fascista*. Florence: A. Vallecchi, 1933.

Allen, Frederick Lewis. *Only Yesterday*. New York: Harper, 1931.

"America in 1960." *Life*, June 5, 1939, 81–84.

Andreotti, Libero. "The Techno-Aesthetics of Shock: Mario Sironi and Italian Fascism." *Grey Room* 38 (2010): 38–61.

Antliff, Mark. "Fascism, Modernism, and Modernity." *Art Bulletin* 84 (2002): 148–69.

Armus, Seth. *French Anti-Americanism: Critical Moments in a Complex History*. Lanham, MD: Lexington Books, 2007.

Arthurs, Joshua. *Excavating Modernity: The Roman Past in Fascist Italy*. Ithaca, NY: Cornell University Press, 2012.

Association of Japan International Exposition. *Grand International Exposition of Japan, March–August 1940: In Celebration of the 2600th Anniversary of the Accession of the First Emperor Jimmu to the Throne, the Purport, Aims and Outline of the Exposition*. Tokyo: Association of Japan International Exposition, 1937.

Bardzko, Dieter. "Allemagne." In *Paris 1937: Cinquantenaire de l'exposition internationale des arts et techniques dans la vie moderne*, edited by Bertrand Lemoine, 134–39. Paris: Institut français d'architecture/Paris Musées, 1987.

Barnhart, Michael A. *Japan Prepares for Total War: The Search for Economic Security, 1919–1941.* Ithaca, NY: Cornell University Press, 1987.

Beauchamp, K.G. *Exhibiting Electricity.* London: Institution of Electrical Engineers, 1997.

Bellassai, Sandro. "The Masculine Mystique: Anti-Modernism and Virility in Fascist Italy." *Journal of Modern Italian Studies* 10, no. 3 (2005): 314–35.

Berghoff, Hartmut. "Enticement and Deprivation: The Regulation of Consumption in Pre-War Nazi Germany." In *The Politics of Consumption: Material Culture and Citizenship in Europe and America*, edited by Martin Daunton and Matthew Hilton, 165–84. Oxford: Berg, 2001.

Bernbach, William, and Herman Jaffe, eds. *Book of Nations: New York World's Fair.* New York: Winkler and Kelmans, 1939.

Bertilaccio, Carlo, and Francesco Innamorati. *EUR, E42 Heritage: A Useful Handbook.* Rome: Palombi, 2005.

Betts, Paul. *The Authority of Everyday Objects.* Berkeley: University of California Press, 2004.

Betts, Paul. "The Nierentisch Nemesis: Organic Design as West German Pop Culture." *German History* 19, no. 2 (2001): 185–217.

Bird, William. "Enterprise and Meaning: Sponsored Film, 1939–1949." *History Today*, December 1989, 24–30.

Bletter, Rosemarie H. "'The Laissez-Fair,' Good Taste and Money Trees." In *Remembering the Future: The New York World's Fair from 1939 to 1964*, edited by Rosemarie Haag Bletter, 119–22. New York: Rizzoli, 1989.

Botton, Alain de. *The Architecture of Happiness.* New York: Vintage, 2008.

Calichman, Richard F., ed. and trans. *Overcoming Modernity: Cultural Identity in Wartime Japan.* New York: Columbia University Press, 2008.

Calichman, Richard F., ed. and trans. *What Is Modernity?: Writings of Takeuchi Yoshimi.* New York: Columbia University Press, 2005.

Calvesi, Maurizio, Enrico Guidoni, and Lux Simonetta. *Ideologia e programma dell'olimpiade della civiltà.* Venice: Marsilio, 1987.

Campbell, Joan. *The German Werkbund: The Politics of Reform in the Applied Arts.* Princeton, NJ: Princeton University Press, 1978.

Carli, Carlo Fabrizio, Gianni Mercurio, and Luigi Prisco. *E42 EUR: Segno e sogno del Novecento*. Rome: DataArs, 2005.

Cassata, Francesco. *Building the New Man: Eugenics, Racial Science and Genetics in Twentieth-Century Italy*. Translated by Erin O'Loughlin. Budapest: Central European Press, 2011.

Castillo, Greg. "Making a Spectacle of Restraint: The Deutschland Pavilion at the 1958 Brussels Exposition." *Journal of Contemporary History* 47 (2012): 97–119.

Cawelti, John. "America on Display: The World's Fairs of 1876, 1893, 1933." In *The Age of Industrialism in America: Essays in Social Structure and Cultural Values*, edited by Frederic Cople Jaher, 317–63. New York: Free Press, 1968.

Cho, Hyunjung. "Hiroshima Peace Memorial Park and the Making of Japanese Postwar Architecture." *Journal of Architectural Education* 66, no. 1 (2012): 72–83.

Ciucci, Giorgio. "The Classicism of the E42: Between Modernity and Tradition." Translated by Jessica Levine. *Assemblage* 8 (1989): 79–87.

Ciucci, Giorgo. "Una prima conclusion: l'E42." In *Gli architetti italiani, architettura e città 1922–1942*, 177–200. Turin: Einaudi, 1989.

Cohen, Jean-Louis. "U.R.S.S." In *Paris 1937: Cinquantenaire de l'exposition internationale des arts et techniques dans la vie moderne*, edited by Bertrand Lemoine, 184–89. Paris: Institut français d'architecture/Paris Musées, 1987.

Commemorative Association for the Japan World Exposition. *Official Report of the Japan World Exposition, Osaka, 1970*. Vol. 1. Suita City, Osaka: Commemorative Association for the Japan World Exposition, 1970.

Comune di Roma, Assessorato alle politiche culturali. *Museo della civiltà romana*. Rome: Comune di Roma, 2008.

Confino, Alon, and Rudy Koshar. "Regimes of Consumer Culture: New Narratives in Twentieth-Century German History." *German History* 19, no. 2 (2001): 135–61.

Corn, Joseph, and Brian Horrigan. *Yesterday's Tomorrows: Past Visions of the American Future*. Baltimore, MD: Johns Hopkins University Press, 1996.

Cotter, Bill. *The 1939–1940 New York World's Fair*. Charleston, SC: Arcadia Publishing, 2009.

Cull, Nicholas. "Overture to an Alliance: British Propaganda at the New York World's Fair 1939–1940." *Journal of British Studies* 36, no. 3 (1997): 325–54.

Cusker, Joseph. "The World of Tomorrow: The 1939 New York World's Fair." PhD diss., Rutgers University, 1990.

Dautry, Roger. "Culture and Technique." *Harvard Business Review* 12 (1934): 409–12.

Day, Charles R. *Education for the Industrial World*. Cambridge: MIT Press, 1987.

Day, Charles R. *Schools and Work: Technical and Vocational Education in France since the Third Republic*. Montreal: McGill-Queen's University Press, 2001.

De Grazia, Victoria. *The Culture of Consent: Mass Organization of Leisure in Fascist Italy*. Cambridge: Cambridge University Press, 1981.

Devos, Rika. "Power, Nationalism and National Representation in Modern Architecture and Exhibition Design at Expo 58." In *Nationalism and Architecture*, edited by Raymond Quek and Darren Deane, 81–94. Burlington, VT: Ashgate, 2012.

Di Majo, Luigi, and Italo Insolera. *L'EUR e Roma dagli anni Trenta al Duemila*. Rome: Editori Laterza, 1986.

Doak, Kevin Michael. *Dreams of Difference: The Japan Romantic School and the Crisis of Modernity*. Berkeley: University of California Press, 1994.

Dolkart, Andrew S. "The Skyscraper City." *The Architecture and Development of New York City*. Edited by Vivian Ducat, Nathanial Herz, Arjun Mehra, et al. Columbian University Digital Knowledge Ventures, 2004. http://ci.columbia.edu/0240s/0242_3/0242_3_fulltext.pdf.

Duffus, R. L. "The Beginning of a World, Not the End." *New York Times Magazine*, July 2, 1939, 1–2.

Duhamel, Georges. *America the Menace*. Translated by Charles Miner Thompson. London: Allen and Unwin, 1931.

DuMond, Jesse W. M. "The 'Palace of Discovery' at the Paris Exposition of 1937." *Journal of Applied Physics* 9, no. 5 (1938): 289–94.

Dupays, Paul. *L'exposition internationale de 1937: Ses creations et ses merveilles*. Paris: H. Didier, 1938.

Duranti, Marco. "Utopia, Nostalgia and World War at the 1939–40 New York World's Fair." *Journal of Contemporary History* 41, no. 4 (2006): 663–83.

Ekirch, Arthur, Jr. *Ideologies and Utopias: The Impact of the New Deal on American Thought*. Chicago: Quadrangle, 1969.

Ellwood, David. *The Shock of America: Europe and the Challenge of the Century*. Oxford: Oxford University Press, 2012.

Elsner, Henry, Jr. *The Technocrats*. Syracuse, NY: Syracuse University Press, 1967.

"Esposizione Universale Romana (E 42)." In *Dizionario del fascismo*, edited by Victoria de Grazia and Sergio Luzzato, 488–90. Turin: Einaudi, 2002.

Etlin, Richard A. *Modernism in Italian Architecture, 1890–1940*. Cambridge: MIT Press, 1991.

Fickers, Andreas. "Presenting the 'Window on the World' to the World: Competing Narratives of the Presentation of Television at the World's Fairs in Paris (1937) and New York (1939)." *Historical Journal of Film, Radio and Television* 28, no. 3 (2008): 291–310.

Fiss, Karen. *Grand Illusion: The Third Reich, the Paris Exposition, and the Cultural Seduction of France*. Chicago: University of Chicago Press, 2010.

Flachowsky, Sören. *Von der Notgemeinschaft zum Reichsforschungsrat: Wissenschaftspolitik im Kontext von Autarkie, Aufrüstung, und Krieg*. Stuttgart: Franz Steiner, 2008.

Fogu, Claudio. *The Historic Imaginary: Politics of History in Fascist Italy*. Toronto: University of Toronto Press, 2003.

Frampton, Kenneth. *Modern Architecture: A Critical History*. 4th ed. New York: Thames and Hudson, 2007.

Friebe, Holm. "Branding Germany: Hans Domizlaff's Markentechnik and Its Ideological Impact." In *Selling Modernity: Advertising in Twentieth-Century Germany*, edited by Pamela E. Swett, S. Jonathan Wiesen, and Jonathan R Zatlin, 78–101. Durham, NC: Duke University Press, 2007.

Fritzsche, Peter. "Nazi Modern." *Modernism/Modernity* 3, no. 1 (1996): 1–22.

Fujihara, Ginjirō. *The Spirit of Japanese Industry*. Tokyo: Hokuseido Press, 1936.

Fuller, Mia. "Wherever You Go, There You Are: Fascist Plans for the Colonial City of Addis Ababa and the Colonizing Suburb of EUR '42." *Journal of Contemporary History* 31, no. 2 (1996): 397–418.

Gagnon, Paul A. "French Views of the Second American Revolution." *French Historical Studies* 2, no. 4 (1962): 430–49.

Galluzzi, Paolo "La storia della scienza nell'E42." In Vol. 1 of *E42: Utopia e Scenario del Regime*, edited by Tullio Gregory, and Achille Tartaro, 53–69. Venice: Cataloghi Marsilio, 1987.

General Motors. *Your Guide to General Motors Highways and Horizons Exhibit New York World's Fair 1939*. New York: General Motors, 1939.

Germer, Andrea. "Artists and Wartime Politics: Natori Yōnosuke—a Japanese Riefenstahl?" *Journal of the German Institute for Japanese Studies* 24 (2012): 21–50.

Germer, Andrea. "Visual Propaganda in Wartime East Asia: The Case of Natori Yōnosuke." *Asia-Pacific Journal* 9, no. 3 (2011). http://www.japanfocus.org/-Andrea-Germer/3530.

Geyer, Michael. "Germany, or, the Twentieth Century as History." *South Atlantic Quarterly* 96, no. 4 (1997): 693–702.

Ghirardo, Diane. "Architects, Exhibitions, and the Politics of Culture in Fascist Italy." *Journal of Architectural Education* 45, no. 2 (1992): 67–75.

Ghirardo, Diane. *Building New Communities: New Deal America and Fascist Italy.* Princeton, NJ: Princeton University Press, 1989.

Ghirardo, Diane. "Città Fascista: Surveillance and Spectacle." In "The Aesthetics of Fascism." Special issue, *Journal of Contemporary History* 31, no. 2 (Apr. 1996): 347–72.

Ghirardo, Diane. "Italian Architects and Fascist Politics: An Evaluation of the Rationalist's Role in Regime Building." *Journal of the Society of Architectural Historians* 39, no. 2 (1980): 109–27.

Greenhalgh, Paul. *Ephemeral Vistas: The Expositions Universelles, Great Exhibitions and World's Fairs, 1851–1939.* Manchester: Manchester University Press, 1988.

Gregory, Tullio, and Achille Tartaro, eds. *E42: Utopia e Scenario del Regime.* 2 vols. Venice: Cataloghi Marsilio, 1987.

Griffin, Roger. *The Nature of Fascism.* New York: St. Martin's, 1991.

Griffin, Roger. "The Palingenetic Core of Fascist Ideology." In *Che cos'è il fascismo? Interpretazioni e prospettive di recherche*, edited by Alessandro Campi, 97–122. Rome: Ideazione, 2003.

Günther, Irene. *Nazi Chic?: Fashioning Women in the Third Reich.* Oxford: Berg, 2004.

Hachmeister, Lutz. "Die Welt des Joseph Goebbels." In *Das Goebbels-Experiment: Propaganda und Politik*, edited by Lutz Hachmeister and Michael Kloft, 7–15. Munich: Deutsche Verlags-Anstalt, 2005.

Haddow, Robert. "Material Culture and the Cold War: International Trade Fairs and the American Pavilion at the 1958 Brussels World's Fair." PhD diss., University of Minnesota, 1994.

Haddow, Robert. *Pavilions of Plenty: Exhibiting American Culture Abroad in the 1950s.* Washington, DC: Smithsonian Institution Press, 1997.

Hagan, Carol. "Visions of the City at the 1939 New York World's Fair." PhD diss., University of Pennsylvania, 2000.

Hart, Jeffrey. "The Last Great Fair." *New Criterion* 23 (January 2005): 74–78.

Hearings before the Committee on Foreign Affairs, House of Representatives, 75th Congress, on H.J. Res. 234, March 23, 1937. Washington, DC: Government Printing Office, 1937.

Herf, Jeffrey. *Reactionary Modernism: Technology, Culture, and Politics in Weimar and the Third Reich.* Cambridge: Cambridge University Press, 1984.

Herf, Jeffrey. "Reactionary Modernism Reconsidered." In *The Intellectual Revolt against Liberal Democracy 1870–1945,* edited by Z. Sternhell, 131–58. Jerusalem: Israel Academy, 1996.

Hildebrand, Winfried Sonia. "Die Ausstellung als Erlebnis—'Gebt mir vier Jahre Zeit!'" In *100 Jahre Deutscher Werkbund 1907–2007,* edited by Winfried Nerdinger, 209–20. Munich: Prestel, 2007.

"A Historical Outline of the Development of Manchurian Railways." *Contemporary Manchuria* 3, no. 3 (July 1939): 36–61.

Hixson, Walter. *Parting the Curtain: Propaganda, Culture and the Cold War, 1945–1961.* New York: St. Martin's, 1997.

Hobsbawm, Eric. "Mass-Producing Traditions: Europe 1870–1914." In *The Invention of Tradition,* edited by Eric Hobsbawm and Terence Ranger, 263–307. Cambridge: Cambridge University Press, 1983.

Hofstadter, Richard. *The Age of Reform: From Bryan to F.D.R.* New York: Knopf, 1955.

Holton, Gerald. *Science and Anti-Science.* Cambridge, MA: Harvard University Press, 1993.

Horiguchi, Sutemi. "'Japanese Taste' in Modern Architecture." Translated by Robin Thompson. *Art in Translation* 4, no. 4 (2012): 407–34.

Huhn, Rosi. "Art et technique: la lumière." In *Paris 1937: Cinquantenaire de l'exposition internationale des arts et techniques dans la vie moderne,* edited by Bertrand Lemoine, 400–401. Paris: Institut français d'architecture/Paris Musées, 1987.

Huizinga, Johan. *In the Shadow of To-Morrow: A Diagnosis of the Spiritual Distemper of Our Time.* London: William Heinemann, 1936.

Iida, Yumiko. *Rethinking Identity in Modern Japan: Nationalism as Aesthetics.* London: Routledge, 2002.

Institut International de Coopération Intellectuelle, ed. *Gendai jin no kensetsu*

[The formation of the modern man]. Translated by Satō Masaaki. Tokyo: Sōgensha, 1937.

"International Exposition of Japan in 1940 to be held at Tokyo and Yokohama to Commemorate 2,600th Year of Founding of Japanese Empire." *Japan Trade Review* 11, no. 2 (March 1938): 37–44.

Irace, Furio. "Reconsidering Architecture: EUR (Rome): 1937–1987." *Abitare* 255 (June 1987): 178–85, 192.

Isozaki, Arata. *Japan-ness in Architecture*. Translated by Sabu Kohso. Edited by David B. Stewart. Cambridge: MIT Press, 2011.

Ivy, Marilyn. "Foreword: Fascism, Yet?" In *The Culture of Japanese Fascism*, edited by Alan Tansman, vii–xii. Durham, NC: Duke University Press, 2009.

Kaji-O'Grady, Sandra. "Authentic Japanese Architecture after Bruno Taut: The Problem of Eclecticism." *Fabrications* 11, no. 2 (2001): 1–12.

Kallis, Aristotle. *The Third Rome, 1922–43: The Making of the Fascist Capital*. New York: Palgrave Macmillan, 2014.

Kargon, Robert, and Arthur Molella. *Invented Edens: Techno-Cities of the Twentieth Century*. Cambridge: MIT Press, 2008.

Kawahata Naomichi. "Pari Bankoku Hakurankai: Roteishita sho mondai" [The Paris International Exposition: an exposé of various issues]. In *"Teikoku" to bijutsu: 1930 nendai Nihon no taigai bijutsu senro* ["Empire" and art: Japanese art of the 1930s and its strategic expansion abroad], edited by Omuka Toshiharu, 405–44. Tokyo: Kokusho Hankōkai, 2010.

Kihlstedt, Folke. "Utopia Realized: The World's Fairs of the 1930s." In *Imagining Tomorrow: History, Technology, and the American Future*, edited by Joseph Corn, 97–118. Cambridge: MIT Press, 1988.

Kohn, Robert. "Social Ideals in a World's Fair." *North American Review* 247, no. 1 (1939): 115–20.

Kostoff, Spiro. *The Third Rome, 1870–1950: Traffic and Glory*. Berkeley, CA: University Art Museum, 1973.

Krenn, Michael. "Unfinished Business: Segregation and U.S. Diplomacy at the 1958 World's Fair." *Diplomatic History* 20, no. 4 (1996): 591–612.

Kushner, Barak. *The Thought War: Japanese Imperial Propaganda*. Honolulu: University of Hawaii Press, 2006.

Kuznick, Peter. "Losing the World of Tomorrow: The Battle over the Presentation of Science at the 1939 New York World's Fair." *American Quarterly* 46 (1994): 341–73.

Labbé, Edmond. *Paris, Exposition Internationale des Arts et Techniques dans la vie Moderne (1937), Rapport Général.* Paris: Imprimerie nationale, 1938–1940.

Labbé, Edmond. *Le régionalisme et l'exposition internationale de Paris 1937.* Paris: Imprimerie nationale, 1936.

Lemoine, Bertrand, ed. *Paris 1937: Cinquantenaire de l'exposition internationale des arts et techniques dans la vie moderne.* Paris: Institut français d'architecture/ Paris Musées, 1987.

Lemoine, Bertrand, and Philippe Rivoirard. "Electricité et Lumière." In *Paris 1937: Cinquantenaire de l'exposition internationale des arts et techniques dans la vie moderne,* edited by Bertrand Lemoine, 222–23. Paris: Institut français d'architecture/Paris Musées, 1987.

Léon, Paul. "Each to His Kind." *Exposition Paris 1937* 12 (May 1937): 15.

Levine, Alison J. Murray. *Framing the Nation: Documentary Film in Interwar France.* New York: Continuum, 2010.

Lewis, Sinclair. *Babbitt.* New York: New American Library, 1961. First published 1922 by Harcourt, Brace and Co.

Lichtenberg, B. "Business Backs New York World Fair to Meet the New Deal Propaganda." *Public Opinion Quarterly* 2, no. 2 (April 1938): 314–20.

Lippmann, Walter. "A Day at the World's Fair." *Current History* 50 (July 1939): 50–51.

Littlejohn, David. "États-Unis." In *Paris 1937: Cinquantenaire de l'exposition internationale des arts et techniques dans la vie moderne,* edited by Bertrand Lemoine, 156–57. Paris: Institut français d'architecture/Paris Musées, 1987.

Lux, Simonetta, and Giorgio Muratore. *Palazzo dei congressi.* Rome: Editalia, 1990

M., J. "Science and the New York World's Fair." *Scientific Monthly* 48 (May 1939): 471–75.

Mack Smith, Denis. *Mussolini.* New York: Knopf, 1982.

Maiwald, E. W. *Reichsausstellung Schaffendes Volk Düsseldorf 1937. Ein Bericht.* Düsseldorf: Bagel, 1939.

"Manchuria's Super-Express 'Asia.'" *Contemporary Manchuria* 2, no.1 (1938): 45–60.

Marchand, Roland. "Corporate Imagery and Popular Education: World's Fairs and Expositions in the United States, 1893–1940." In *Consumption and American Culture,* edited by David Nye and Carl Pedersen, 18–33. Amsterdam: VU University Press, 1991.

Marchand, Roland. *Creating the Corporate Soul: The Rise of Public Relations and Corporate Imagery in American Big Business.* Berkeley: University of California Press, 1998.

Marchand, Roland. "The Designers Go to the Fair II: Norman Bel Geddes, The General Motors 'Futurama' and the Visit to the Factory Transformed." *Design Issues* 8, no. 2 (1992): 22–40.

Marefat, Mina. "Building to Power: Architecture of Tehran, 1921–1941." PhD diss., MIT, 1988.

Mariani, Riccardo, ed. *E 42: Un progetto per "l'Ordine Nuovo."* Milan: Edizioni Comunità, 1987.

Mariani, Riccardo. "La progettazione dell'E42, prima fase." *Lotus* 67 (1990): 90–126.

Marx, Karl, and Friedrich Engels. *Basic Writings on Politics and Philosophy.* Edited by Lewis Feuer. Garden City, NY: Anchor Books, 1959.

Masumoto, Naofumi. "Interpretations of the Filmed Body: An Analysis of the Japanese Version of Leni Riefenstahl's *Olympia.*" In *Critical Reflections on Olympic Ideology: Second International Symposium for Olympic Research,* edited by Robert K. Barney and Klaus V. Meier, 146–57. London, Ontario: Centre for Olympic Studies, University of Western Ontario, 1994.

Mathy, Jean-Philippe. *Extreme-Occident: French Intellectuals and America.* Chicago: University of Chicago Press, 1993.

Mauro, James. *Twilight at the World of Tomorrow: Genius, Madness, Murder, and the 1939 World's Fair on the Brink of War.* New York: Ballantine Books, 2010.

McCloskey, Barbara. *Artists of World War II.* Westport, CT: Greenwood Press, 2005.

McNeil, Peter. "Myths of Modernism: Japanese Architecture, Interior Design and the West, c. 1920–1940." *Journal of Design History* 5, no. 4 (1992): 281–94.

Mirzoeff, Nicholas. *An Introduction to Visual Culture.* London: Routledge, 1999.

Mizuno, Hiromi. *Science for the Empire: Scientific Nationalism in Modern Japan.* Stanford, CA: Stanford University Press, 2009.

Moholy-Nagy, Laszlo. "The 1937 International Exhibition, Paris." *Architectural Record* 82 (October 1937): 82.

Mumford, Lewis. *Sidewalk Critic.* Edited by Robert Wojtowicz. New York: Princeton Architectural Press, 1998.

Muntoni, Alessandra. "Esposizioni e regime, ideologie e colonialism, e Roma 1942, L'Esposizione Universale." *Quaderni DI* 11 (1990): 61–70, 175–80.

Muntoni, Alessandra. "La Vicenda dell'E42. Fondazione di una città in forma didascalica." In *Classicismo/Classicismi, Quaderni di Architettura*, edited by Giorgio Ciucci, 128–43. Milan: Electa, 1995.

Muratore, Giorgio. "Die Überwindung des ersten Modernismus: Eine neue Stadt für die Weltausstellung 1942-E'42." In Vol. 2 of *Kunst und Diktatur: Architektur, Gilderhauerei und Malerei in Österreich, Deutschland, Italien und der Sowjetunion 1922–1956*, edited by Jan Tabor, 632–37. Baden: Grasl, 1994.

Najita, Tetsuo, and H. D. Harootunian. "Japanese Revolt against the West: Political and Cultural Criticism in the Twentieth Century." In *The Cambridge History of Japan*. Vol. 6, *The Twentieth Century*, edited by Peter Duus, 711–74. Cambridge: Cambridge University Press.

Nelis, Jan. "Constructing Fascist Identity: Benito Mussolini and the Myth of 'Romanità.'" *Classical World* 100, no. 4 (2007): 391–415.

Nolan, Mary. "America in the German Imagination." In *Transactions, Transgressions, Transformations: American Culture in Western Europe and Japan*, edited by Heide Fehrenbach and Uta Poiger, 3–25. New York: Berghahn Books, 2000.

Nolan, Mary. *Visions of Modernity: American Business and the Modernization of Germany*. New York: Oxford University Press, 1994.

Notaro, Anna. "Exhibiting the New Mussolinian City: Memories of Empire in the World Exhibition of Rome (EUR)." *Geojournal* 51, no. 1/2 (2001): 15–22.

O'Connor, F. V. "The Usable Future: The Role of Fantasy in the Promotion of a Consumer Society for Art." In *Dawn of a New Day: The New York World's Fair 1939/40*, edited by Helen Harrison and Joseph Cusker, 57–71. New York: NYU Press, 1980.

Ory, Pascal. *La Belle Illusion: Culture et politique sous le signe du Front Populaire 1935–1938*. Paris: Plon, 1994.

Ory, Pascal. "Le Front Populaire et l'Exposition." In *Paris 1937: Cinquantenaire de l'exposition internationale des arts et techniques dans la vie moderne*, edited by Bertrand Lemoine, 30–35. Paris: Institut français d'architecture/Paris Musées, 1987.

Painter, Borden W., Jr. *Mussolini's Rome: Rebuilding the Eternal City*. New York: Palgrave Macmillan, 2005.

Panchasi, Roxanne. *Future Tense: The Culture of Anticipation in France between the Wars*. Ithaca, NY: Cornell University Press, 2009.

Peer, Shanny. *France on Display: Peasants, Provincials and Folklore in the 1937 Paris World's Fair*. Albany: SUNY Press, 1998.

Peer, Shanny. "Radio." In *Paris 1937: Cinquantenaire de l'exposition internationale des arts et techniques dans la vie moderne*, edited by Bertrand Lemoine, 242–43. Paris: Institut français d'architecture/Paris Musées, 1987.

Peretz, Don. *The Middle East Today*. 6th ed. Westport, CT: Praeger, 1994.

Perrin, Jean. "The Palace of Discovery." *Exposition Paris 1937* 2 (June 1936): 5.

Perrin, Jean. "The Palace of Discovery." *Exposition Paris 1937* 12 (May 1937): 6–9.

Petropoulos, Jonathan. *Royals and the Reich: The Princes von Hessen in Nazi Germany*. Oxford: Oxford University Press, 2006.

Petsch, Joachim. "Möbeldesign im Dritten Reich und die Erneuerung des Tischler-Gewerbes seit dem ausgehenden 19. Jahrundert." In *Design in Deutschland 1933–45. Ästhetik und Organisation des Deutschen Werkbundes im "Dritten Reich,"* edited by Sabine Weißler, 42–55. Berlin: Giessen/Werkbund Archiv, 1990.

Plum, Gilles. "Chemins de Fer." In *Paris 1937: Cinquantenaire de l'exposition internationale des arts et techniques dans la vie moderne*, edited by Bertrand Lemoine, 218–19. Paris: Institut français d'architecture/Paris Musées, 1987.

Plum, Gilles. "Le Palais de la Découverte et le Grand Palais." In *Paris 1937: Cinquantenaire de l'exposition internationale des arts et techniques dans la vie moderne*, edited by Bertrand Lemoine, 294–99. Paris: Institut français d'architecture/ Paris Musées, 1987.

President's Science Advisory Committee. *Introduction to Outer Space*. Washington, DC: Government Printing Office, 1958.

Quilici, Vieri. "L'E42 in evidenza." *Urbanistica* 88 (1987): 68–84.

Quilici, Vieri. *E42-EUR un centro per la metropolis*. Rome: Photoatlante, 1996.

Ragot, Gilles. "Le Corbusier et l'Exposition." In *Paris 1937: Cinquantenaire de l'exposition internationale des arts et techniques dans la vie moderne*, edited by Bertrand Lemoine, 72–79. Paris: Institut français d'architecture/Paris Musées, 1987.

Reagin, Nancy R. *Sweeping the German Nation: Domesticity and National Identity in Germany, 1870–1945*. New York: Cambridge University Press, 2007.

Reaven, Sheldon J. "New Frontiers: Science and Technology at the Fair." In *Remembering the Future: The New York World's Fair from 1939 to 1964*, edited by Rosemarie Haag Bletter, 75–105. New York: Rizzoli, 1989.

Rébérioux, Madeleine. "L'exposition de 1937 et le contexte politique des années trente." In *Paris 1937: Cinquantenaire de l'exposition internationale des arts et techniques dans la vie moderne*, edited by Bertrand Lemoine, 26–29. Paris: Institut français d'architecture/Paris Musées, 1987.

Reid, Susan. "The Soviet Pavilion at Brussels '58: Convergence, Conversion, Critical Asssimilation or Transculturism?" Cold War History Project, Working Paper no. 62, Woodrow Wilson Center for Scholars, December 2010.

Reynolds, Jonathan M. "Ise Shrine and a Modernist Construction of Japanese Tradition." *Art Bulletin* 83, no. 2 (June 2001): 316–41.

Reynolds, Jonathan M. *Maekawa Kunio and the Emergence of Japanese Modernist Architecture*. Berkeley: University of California Press, 2001.

Reynolds, Jonathan M. "Teaching Architectural History in Japan: Building a Context for Contemporary Practice." *Journal of the Society of Architectural Historians* 61, no. 4 (2002): 530–36.

Roche, Maurice. *Mega-Events and Modernity: Olympics and Expos in the Growth of Global Culture*. London: Routledge, 2000.

Ross, Corey. "Visions of Prosperity: The Americanization of Advertising in Interwar Germany." In *Selling Modernity: Advertising in Twentieth-Century Germany*, edited by Pamela E. Swett, S. Jonathan Wiesen, and Jonathan R Zatlin, 52–77. Durham, NC: Duke University Press, 2007.

Rydell, Robert. "The Fan Dance of Science: American World's Fairs in the Great Depression." *Isis* (1985): 525–42.

Rydell, Robert. *World of Fairs: The Century-of-Progress Expositions*. Chicago: University of Chicago Press, 1993.

Schäfers, Stefanie. *Vom Werkbund zum Vierjahresplan: Die Ausstellung 'Schaffendes Volk,' Düsseldorf 1937*. Düsseldorf: Droste, 2001.

Schlesinger, Arthur M., Jr. *The Age of Roosevelt*. Vol. 2, *The Coming of the New Deal*. Boston: Houghton Mifflin, 1960.

Schnapp, Jeffrey T. "Epic Demonstrations: Fascist Modernity and the 1932 Exhibition of the Fascist Revolution." In *Fascism, Aesthetics, and Culture*, edited by Richard J. Golsan, 2–77. Hanover, NH: University Press of New England, 1992.

Schnapp, Jeffrey. "Fascism's Museum in Motion." *Journal of Architectural Education* 45, no. 2 (1992): 87–97.

Schnapp, Jeffrey. "Flash Memories (Sironi on Exhibit)." *South Central Review* 21, no. 1 (2004): 22–49.

Schroeder-Gudehus, Brigitte, and Anne Rasmussen. *Les Fastes du Progrès: Le guide des Expositions universelles 1851–1992*. Paris: Flammarion, 1992.

Schutts, Jeff. "'Die erfrischende Pause': Marketing Coca-Cola in Hitler's Germany." In *Selling Modernity: Advertising in Twentieth-Century Germany*, edited by Pamela E. Swett, S. Jonathan Wiesen, and Jonathan R Zatlin, 151–81. Durham, NC: Duke University Press, 2007.

Schwantes, Robert S. "Japan's Cultural Foreign Policies." In *Japan's Foreign Policy, 1868–1941: A Research Guide*, edited by James W. Morley, 153–83. New York: Columbia University Press, 1974.

Scully, Vincent. "Louis I. Kahn and the Ruins of Rome." *Museum of Modern Art Members Quarterly*, summer 1992, 1–13.

Seidensticker, Edward. *Tokyo Rising: The City since the Great Earthquake*. New York: Alfred A. Knopf, 1990.

Shibasaki, Atsushi. "Activities and Discourses on International Cultural Relations in Modern Japan: The Making of KBS (Kokusai Bunka Shinko Kai), 1934–53." *Journal of Global Media Studies* 8 (March 2011): 25–41.

Shin, Mizukoshi. "Social Imagination and Industrial Formation of Television in Japan." *Bulletin of the Institute of Socio-Information and Communication Studies, University of Tokyo* (March 1996): 1–14.

Shinanyaku. *Moboroshi no 1940 nen keikaku: Taiheiyō sensō no zenya, 'kiseki no toshi' ga tanjō shita* [The dream plan of 1940: the birth of the "City of Wonder" just before the Pacific War]. Tokyo: Asupekuto, 2009.

"Shinsō nareru banpaku jimukyoku o rikugun shōbyōhei shūyōjo ni katsuyō" [Refurbished expo headquarters put to practical use as a care facility for sick and wounded soldiers]. *Banpaku* 30 (November 1938): 4–5.

Siegelbaum, Lewis. "Sputnik Goes to Brussels: The Exhibition of a Soviet Technological Wonder." *Journal of Contemporary History* 47, no. 1 (2012): 120–36.

Siegfried, André. *America Comes of Age: A French Analysis*. New York: Harcourt and Brace, 1927.

Smith, Anthony D. *Nationalism and Modernism: A Critical Survey of Recent Theories of Nations and Nationalism*. London: Routledge, 1998.

Snider, Sage. "Realizing the Fascist Vision: Mussolini's Construction of Roman History at EUR." Undergraduate thesis, Yale University, 2012.

Spoerri, William. *The Old World and the New: A Synopsis of Current European Views on American Civilization*. Antwerpen: Imprimerie du Centre, 1936.

Stewart, David B. *The Making of a Modern Japanese Architecture: From the Founders to Shinohara and Isozaki.* Tokyo: Kodansha International, 2002.

Stone, Marla. "A Fascist Theme Park." In *Visual Sense: A Cultural Reader,* edited by Elizabeth Edwards and Kaushik Bhaumik, 271–80. New York: Berg, 2008.

Stone, Marla. "Staging Fascism: The Exhibition of the Fascist Revolution." *Journal of Contemporary History* 28 (April 1993): 215–43.

Strang, G. Bruce. "'The Worst of all Worlds': Oil Sanctions and Italy's Invasion of Abyssinia, 1935–1936." *Diplomacy and Statecraft* 19, no. 2 (2008): 210–35.

Stratigakos, Despina. "What Is a German Home? Interior Domestic Design and National Identity in the Third Reich." Paper presented at the annual meeting of the College Art Association, Los Angeles, February 2012.

Strauss, David. *Menace in the West: The Rise of French Anti-Americanism in Modern Times.* Wesport, CT: Greenwood, 1978.

Suga, Yasuko. "Modernism, Nationalism and Gender: Crafting 'Modern' Japonisme." *Journal of Design History* 21, no. 3 (2008): 259–75.

Suganami, Sohji. "Japan." *Exposition Paris 1937: Arts, Crafts, Sciences in Modern Life* 14 (July–August 1937): 14–15.

Swett, Pamela E., S. Jonathan Wiesen, and Jonathan R. Zatlin. "Introduction." In *Selling Modernity: Advertising in Twentieth-Century Germany,* 1–26. Durham, NC: Duke University Press, 2007.

Swift, Anthony. "The Soviet World of Tomorrow at the New York World's Fair, 1939." *Russian Review* 57, no. 3 (1998): 364–79.

Tafuri, Manfredo. *La vicenda dell'E42.* In *Ludovico Quaroni e lo sviluppo dell'architettura moderna in Italia,* edited by Manfredo Tafuri, 51–67. Milan: Comunità, 1964.

Takenaka, Akiko. "The Construction of a Wartime National Identity: The Japanese Pavilion at New York World's Fair 1939/40." Master's thesis, MIT, 1997.

Tange, Kenzō. "Dai tōa kyōeiken ni okeru kaiin no yōbō" [Member's plea in regard to the Greater East Asia Co-prosperity Sphere]. *Kenchiku zasshi* [Architectural journal] 56, no. 690 (September 1942): 744.

Tange, Kenzō, Kawazoe Noboru, and Watanabe Yoshio. *Ise: Nihon kenchiku no genkei.* Tokyo: Asahi Shinbun, 1962.

Tange, Kenzō, Noboru Kawazoe, and Yoshio Watanabe. *Ise: Prototype of Japanese*

Architecture. Translated by Eric Klestadt and John Bester. Cambridge: MIT Press, 1965.

Tansman, Alan. "Introduction: The Culture of Japanese Fascism." In *The Culture of Japanese Fascism*, 1–28. Durham, NC: Duke University Press, 2009.

Taut, Bruno. *Fundamentals of Japanese Architecture.* Tokyo: Kokusai Bunka Shinkokai, 1936.

Thamer, Hans-Ulrich. "Geschichte und Propaganda. Kulturhistorische Ausstellungen in der NS-Zeit." *Geschichte und Gesellschaft* 24, no. 3 (1998): 349–81.

This Is America: Official United States Guide Book Brussels World's Fair 1958. N.p.: Office of the Commissioner General, 1958.

Todd, Frank Morton. *The Story of the Exposition: Being the Official History of the International Celebration Held at San Francisco in 1915.* New York: Putnam, 1921.

Tooze, Adam. *The Wages of Destruction: The Making and Breaking of the Nazi Economy.* New York: Viking, 2007.

Toyosawa, Toyoo. *Sōzō no min, Nihon minzoku* [People of creativity, the Japanese people]. Tokyo: Seinen Shobō, 1941.

Tyng, Francis Edmonds. *Making a World's Fair.* New York: Vantage, 1958.

Udovicki-Selb, Danilo. "Facing Hitler's Pavilion: The Uses of Modernity in the Soviet Pavilion at the 1937 Paris International Exhibition." *Journal of Contemporary History* 47, no. 1 (2012): 13–47.

Valéry, Paul. "The European Spirit." In *History and Politics.* Vol. 10 of *The Collected Works of Paul Valéry,* translated by Denise Folliot and Jackson Mathews, 326–28. London: Routledge and Kegan Paul, 1963.

Valéry, Paul, et al. *Seishin no shōrai: Yōroppa seishin no shōrai* [The future of the spirit: the future of the European spirit]. Translated by Satō Masaaki. Tokyo: Shiba Shoten, 1936.

Watanabe, Hiroshi. *The Architecture of Tokyo: An Architectural History in 571 Individual Presentations.* Stuttgart: Edition Axel Menges, 2001.

Weisen, Jonathan. *Creating the Nazi Marketplace: Commerce and Consumption in the Third Reich.* Cambridge: Cambridge University Press, 2011.

Weisenfeld, Gennifer. "The Expanding Arts of the Interwar Period." In *Since Meiji: Perspectives on the Japanese Visual Arts, 1868–2000,* edited by J. Thomas Rimer with translations by Toshiko McCallum, 66–98. Honolulu: University of Hawaii Press, 2012.

Weisenfeld, Gennifer. "Publicity and Propaganda in 1930s Japan: Modernism as Method." *Design Issues* 25, no. 4 (2009): 13–28.

Weisenfeld, Gennifer. "Touring Japan-as-Museum: *NIPPON* and Other Japanese Imperialist Travelogues." *Positions* 8, no. 3 (2000): 747–93.

Wesemael, Pieter van. *Architecture of Instruction and Delight*. Rotterdam: Uitgeverij 010, 2001.

Wesseling, H. L. "From Cultural Historian to Cultural Critic: Johan Huizinga and the Spirit of the 1930s." *European Review* 10, no. 4 (2002): 485–99.

Whalen, Grover. *Mr. New York: The Autobiography of Grover A. Whalen*. New York: Putnam, 1955.

White, E. B. "The World of Tomorrow." In *Essays of E. B. White*, 111–17. New York: Harper and Row, 1977.

Wilson, Sarah. "The Soviet Pavilion in Paris." In *Art of the Soviets*, edited by Matthew Bown and Brandon Taylor, 106–19. Manchester, UK: Manchester University Press, 1993.

Yamawaki, Iwao. "Reminiscences of Dessau." *Design Issues* 2, no. 2 (1985): 56–68.

Yang, Daqing. *Technology of Empire: Telecommunications and Japanese Expansion in Asia, 1883–1945*. Cambridge, MA: Harvard University Asia Center, 2010.

Yoshida, Mitsukuni. *Bankoku hakurankai: Gijutsu benmei shiteki ni* [International expositions: from a history of technological civilization perspective]. Tokyo: Nihon Hōsō Kyōkai, 1985.

Yoshimi, Shunya. *Hakurankai no seijigaku* [The politics of expositions]. Tokyo: Chūō Kōronsha, 1992.

Your World of Tomorrow. New York: The Fair, 1939.

INDEX

European Economic Community (EEC), 140

Exhibition of the Fascist Revolution (Mostra della rivoluzione fascista), 112, 125–26

Exposition internationale 1937 in Paris, 1, 7–29, 33, 35, 39, 41–42, 71, 123, 139

Exposition Paris 1937: Arts, Crafts, Sciences in Modern Life (magazine), 11, 13, 25

Fahrenkamp, Emil, 38

Fair of the Future Committee, 62, 64

Fascism, 2, 18, 22, 23, 104, 108, 109, 113, 114, 119, 125, 135, 136

Fascist City of the Future (EUR), 115

Fascist: culture, 109, 111, 125, 126, 134; modernism, 111, 116–18, 132; new towns (Foundation Cities), 115–16; planners of EUR, 108; Revolution, 108, 111–12, 125; Superman, 114

Fascist Work 23

Fenderl, Ettore, 121

Fermi, Enrico, 109, 134

Fermi, Laura, 109

Festival of Light, 12, 16

Fiat, 126, 127

Figini, Luigi, 117–18

First World War. *See* World War I

Flushing Meadows, 61, 81

Ford Motor Company, 63

Fordist mass production, 2, 9

Four-Year Plan, 30, 35, 42–44, 46–47, 51–52

France, 1, 6, 7–29, 69, 139

France on Display, 10

Futurama, 73,7 5–77, 79

Futurama II, 81

Futurists, 119

Gagnon, Roger, 9

Galileo, 109, 130–32

Gary, Indiana, 75

GAZ (Soviet Ford), 19

Geddes, Patrick, 64

General Confederation of Labor (CGT), 13

General Electric Company, 63

General Motors Company, 63, 73, 75, 76, 78, 81

Génie du Fascisme, 22

Germany, 1, 2, 5, 7, 20, 21, 24, 69, 138, 141; colonies of, 43, 47; science in, 43–46; Women's Bureau (Deutsches Frauenwerk), 51

Giordani, Francesco, 130, 132

Goebbels, Joseph, 55

Goodman, Benny, 146

Gori, Georges, 22

Göring, Hermann, 35, 44, 45

Grand International Exposition of Japan 1940, 83–93, 99, 104, 107

Graubner, Gerhard, 44

Great Britain (United Kingdom), 2, 5, 69, 139–41

Great Depression, 5, 60, 63

Greater East Asia Co-Prosperity Sphere, 92–93, 95, 97, 100

Great Exhibition (London 1851), 4

Greece, 11

Greenhalgh, Paul, 3

Gropius, Walter, 39

Index   205